How to Excel in Your Doctoral Viva

Stacey Bedwell • Isabelle Butcher

How to Excel in Your Doctoral Viva

palgrave
macmillan

Stacey Bedwell
Institute of Psychiatry, Psychology and
Neuroscience
King's College London
London, UK

Isabelle Butcher
Department of Psychiatry
University of Oxford
Oxford, UK

ISBN 978-3-031-10171-7   ISBN 978-3-031-10172-4   (eBook)
https://doi.org/10.1007/978-3-031-10172-4

Cover Illustration by Firsik

This Palgrave Macmillan imprint is published by the registered company Springer Nature Switzerland AG.
The registered company address is: Gewerbestrasse 11, 6330 Cham, Switzerland

The views and experiences described throughout the book are not affiliated with or endorsed by any institution to which the authors are affiliated.

Preface

The aim of this book is to help those students who are preparing for their PhD or doctoral viva, to feel confident and prepared for the final step in their doctoral journey.

The main purpose of this book is to encourage students with the confidence to reach and achieve their doctoral ward. To achieve this aim, this book is split into 11 key chapters:

- Defining what is the viva
- Common urban myths concerning the viva
- Real viva experiences – reflections from PhD students and reflections from viva examiners
- Making the most of and enjoying your viva
- Possible outcomes as the result of completing your PhD
- Practice questions
- How to own your thesis
- Viva preparation timeline
- Mock viva
- Overcoming viva concerns and anxiety

One of the central aspects of this book is that more than 25 individuals from a range of academic disciplines have kindly contributed their viva experiences. It is hoped that this aspect of the book will enable you, the reader, to understand that each individual's preparation timeline and viva experience is unique.

In this book direct reference is made to the student as 'you'. This is a style which people who had commented on drafts of the chapters valued, as often the doctoral viva experience is an unknown situation so to personalise this, the students are intermittently referred to as 'you'.

Throughout the book, some terms are used interchangeably. For example, we use the term doctoral college/doctoral office/graduate school to describe the administrative office within a higher education institution that supports doctoral students. Where different terms are used to describe the same thing, this reflects the variation in terminology used across the higher education sector.

Stacey A. Bedwell
Isabelle L. Butcher

Acknowledgements

The authors wish to thank their own PhD supervisors, who encouraged us in our own journeys.

The authors also wish to thank all those individuals who kindly contributed to this book by sharing their experiences of vivas, whether as a student or as an examiner.

We are fortunate that colleagues from a range of backgrounds and careers shared their experiences openly and graciously. These contributions make the content of this book far richer.

Finally, thank you to our families for supporting us in our own PhD journeys, and for encouraging us.

Contents

List of Tables

1

Introduction

Stacey Bedwell and Isabelle Butcher

1.1 Welcome

Welcome to 'How to excel in your doctoral viva'. We are very excited to bring you our unique aid to the final phase of your PhD or doctoral journey. The following chapters have been compiled by Dr. Stacey Bedwell and Dr. Isabelle Butcher with the specific aim of guiding you through the process of preparing for your doctoral viva, whether it is coming up very soon or if you are just starting out on your PhD journey and want to be prepared for what is coming.

Throughout the book you will work your way through topics including: what the viva is and what it involves; debunking urban myths surrounding the doctoral viva; recollections from PhD graduates and examiners of their own viva experiences and tips; what you can do to enjoy your own viva and get the most out of the process; a guide to the possible viva outcomes and what they mean; some common viva questions and how to prepare for them; how to own your thesis; and the benefits of participating in a mock viva. As you progress through the book you will be working on a range of skills, thinking points and activities, designed to complement everything you have already developed as a researcher through the course of your doctoral studies. You have already done most of the hard work—you have completed, or nearly

© The Author(s), under exclusive license to Springer Nature Switzerland AG 2022
S. Bedwell, I. Butcher, *How to Excel in Your Doctoral Viva*,
https://doi.org/10.1007/978-3-031-10172-4_1

completed, a doctoral thesis. Each chapter intends to equip you for a successful and enjoyable viva experience and together they will set you on a path towards a positive PhD viva.

1.2   The Authors

Dr Stacey Bedwell is a neurobiologist by background. She was awarded her PhD in neuroanatomy from Nottingham Trent University in 2015 for her work on the anatomical connectivity of the mammalian prefrontal cortex. Prior to her PhD research, Stacey completed an MSc in Clinical Psychology and BSc in Psychology, both at Bangor University.

Since completing her doctoral research, Stacey has completed two postdoctoral research projects at Nottingham Trent University, in neuroanatomy and neurophysiology, before moving to a post as a lecturer in psychology at Birmingham City University. During her time at Birmingham City University, Stacey developed undergraduate modules in research skills and methods, as well as leading the foundation-level psychology course. In 2021 Stacey moved to King's College London to take up a teaching post in psychology in the Institute of Psychiatry, Psychology and Neuroscience. Alongside her full-time role, Stacey is a regular contributor to teaching at the Institute of Continuing Education at the University of Cambridge.

Notable achievements in Stacey's academic career to date include time spent with the House of Commons Science and Technology Committee as part of The Royal Society Pairing Scheme in 2020, and receipt of the Cognitive Neuroscience Society Postdoctoral Fellow Award in 2018. She was shortlisted for Inspirational Teacher of the Year (2021) and awarded second place for Researcher of the Year (2020) by Birmingham City University. During her

PhD Stacey won first prize for presentation of her research at the Nottingham Trent University School of Science and Technology annual conference. Stacey has published in internationally recognised academic journals, including the *European Journal of Neuroscience, Frontiers, Perception* and *Development & Psychopathology*. She has also written for *The Psychologist, Neuroscience News*, BBC and *Scientifica*, as well as appearing on local radio and in national media outlets.

Stacey's current research interests are in the development of high-order cognitive processes associated with the prefrontal cortex, specifically executive functions. Recent projects have investigated the role of childhood experiences of sibling aggression in decision-making manifested in adulthood, the relationship between childhood trauma, psychopathy and response inhibition and the relationship between childhood aggression, violent media, and executive function in children. Ongoing projects are investigating the roles of post-traumatic growth and emotion regulation in executive function and exploring functional network models of decision-making.

Dr. Isabelle Butcher has worked in mental health research in a range of settings and role for several years. More recently, in 2021 Isabelle completed her PhD at the University of Manchester. This explored the association between negative symptoms, traumatic life events and attachment style. Isabelle's PhD was funded by the Medical Research Council. During Isabelle's PhD Isabelle was a member of the European Network of Negative Symptoms (EuroNES) and fostered an interest in the negative symptoms of schizophrenia. Prior to Isabelle's PhD research, Isabelle worked in a range of settings, including inpatient psychiatric clinical settings and a range of universities in health and clinical psychology.

Prior to Isabelle's PhD, Isabelle completed her MSc in Clinical Psychology at Leiden University in the Netherlands and a BSc Psychology at Cardiff

University. As part of Isabelle's undergraduate degree, she completed a professional placement year at the University of Manchester in the Division of Clinical Psychology, where she developed an interest in psychosis.

Since completing her PhD, Isabelle has had the opportunity to work as a project manager and a post-doctoral researcher on a project exploring staff wellbeing in paediatric critical care settings based at a large hospital in central England. In 2022, Isabelle moved to the University of Oxford, working as a postdoctoral researcher on a large multi-site UKRI-funded project exploring adolescent mental health and creative arts in the Department of Psychiatry.

Alongside this role over the last six years Isabelle has cultivated an interest in the ethics of conducting research studies and sits on a NHS research ethics committee as chair. Isabelle has a keen interest in the impact of traumatic life events on mental health and also on the negative symptoms of schizophrenia.

Dr. Stacey Bedwell and Dr. Isabelle Butcher first met when they were both research associates in the same institution and have collaborated on several successful projects over recent years. These include a children's book, *Anything you can do, I can do,* workshops aimed at early career researchers with the British Academy, children's brain workshops with the British Neuroscience Association and research surrounding the psychological phenomenon known as Alice in Wonderland Syndrome.

Throughout some of the following chapters, you will come across some input from additional authors. In specific areas, we have sought the opinions and experiences of a wide range of academics, to give you a comprehensive and truthful insight into the viva experience. We have included the memories and experiences of recent PhD graduates right through to professors who have examined more than 50 PhD theses. The experiences you will read range from medicine, to mathematics, to social sciences. Some experiences are very positive, and some are more negative. Every experience we have included is the unedited words of the original author. We hope the inclusion of these words, from both the student and examiner perspective, help you to develop a clear picture of the viva process for yourself.

1.3 Why Write a Book on how to Complete a Successful Viva?

The doctoral viva has always been somewhat of a mystery and a process that, in comparison to other academic milestones, candidates have historically received very little training and guidance for. The notion of the unknown makes the experience somewhat anxiety-provoking for many (Carleton, 2016).

Research evidence shows that reducing the unknown elements of a situation reduces anxiety (Grupe & Nitschke, 2013). It is reasonable to presume, therefore, that reducing the unknown related to the PhD viva is likely to alleviate some of the viva-related anxiety and fear many candidates experience, allowing them to concentrate more on preparation and enjoying the process.

The idea for this book first came about when the authors were preparing for their own PhD viva exams. There was very limited guidance available on the market that offered any kind of personal insight. We believe this is part of the reason for so many rumours and urban myths circulating postgraduate offices when it comes to the mysterious viva. The mystery is only heightened when you are faced with an exam format that you have never had to do before and for most of us, in the UK at least, cannot watch taking place. This book hopes to give readers an insider's view to the viva and the processes surrounding it, and to lift some of the shroud that has been surrounding the feared academic milestone for so long.

1.4 How Does this Book Work?

The contents of each chapter is designed to be applicable to you regardless of your PhD topic or discipline. We have intentionally avoided focussing on any one discipline throughout. Although there may be places where we draw on examples from specific research fields, or our own experiences, we have aimed to do this in a way that applies regardless of the background you are coming to us from.

We start out in our path to you excelling in your viva by offering a clear explanation of what the viva is, so you can continue with a clear image in your mind of what you are heading towards and why. We then work through each chapter, uncovering the mysteries of the viva experience from several angles. Although we have written each chapter to appear in a certain order, we have also formatted each in such a way as to ensure that they can also function independently. Do read the book cover to cover if that is the approach that best suits you, but we also intend that you can pick up whichever chapter is relevant to you at a given time. You might find that certain topics become more relevant to you as you reach various milestones in the build-up to your viva.

References

Carleton, R. N. (2016). Fear of the unknown: One fear to rule them all? *Journal of Anxiety Disorders, 39,* 30–43. https://doi.org/10.1016/j.janxdis.2016.02.007

Grupe, D. W., & Nitschke, J. B. (2013). Uncertainty and anticipation in anxiety. *Nature Reviews Neuroscience, 14*(7), 488–501. https://doi.org/10.1038/nrn3524

2

What Is the Viva?

Isabelle Butcher

The word viva comes from the Latin 'viva voce', meaning 'living voice.'

The viva is an oral examination and is an opportunity for you as the PhD student to discuss your PhD with two experts, external to your PhD supervisory team, in an engaging manner.

One aspect that makes the viva a mystery is that often each experience differs person to person, as well as higher education institution (HEI) to HEI and thus it is difficult to give guidelines or a handbook of the exact set up. The process of the viva, as we will see later in this chapter and throughout this book, varies and there is no handbook; what can be explained however, is the process and we can offer some guidance as to the background and the setting to try and eradicate some of the myths concerning the doctoral viva examination.

2.1 What the Viva Is Not

The viva, whilst an oral examination, is unlike any other exam, job interview or language oral examination that you may have previously experienced.

© The Author(s), under exclusive license to Springer Nature Switzerland AG 2022
S. Bedwell, I. Butcher, *How to Excel in Your Doctoral Viva*,
https://doi.org/10.1007/978-3-031-10172-4_2

2.2 Purpose of the Viva

The aim of the viva is for you, as the student, to have the opportunity to share your findings in an environment with other academics. As stipulated above, it is often incorrectly assumed that the viva is similar to a job interview. However, this is not an accurate depiction of the viva, the viva is a two-way conversation, a discussion between you as the student and two academics, who are often, as explained, experts in your PhD areas.

2.3 What Does the Viva Involve?

The viva is predominantly a discussion between you and two senior academics. It usually involves between two and four hours of conversation, centred around the contents of your thesis. During the viva, your examiners will ask you questions about the content of each of your chapters, often in a systematic way, from beginning to end. We will cover possible viva questions in more detail in later chapters.

The viva involves some key aspects:

- A discussion between you and two academics about your work.
- Being able to independently articulate your doctoral work.
- Being able to defend your work and the decisions made.
- Being able to teach others about your doctoral work.

2.4 Why Is the Viva an Oral Examination Not a Written Exam?

It is crucial that as a doctoral student you can articulate your findings verbally as well as write them. Once you have submitted your doctoral thesis, you have successfully written your findings. As academics, it is important that you are able to express your work and findings both verbally and through writing. Whilst being able to write is a key component of being an academic, it is not the only skill; verbally sharing your findings is a skill that will enable you to articulate your findings succinctly to other academics but perhaps more importantly to other individuals such as, but not limited to, lay members of the public, other academics, policy makers and charitable organisations. The doctoral viva is a great experience in verbally sharing scientific findings with other academics who may be in your field but who are not experts in your specific research.

The viva is also an oral examination, not a written exam, to ensure that the examiners are confident that you conducted the doctoral work yourself. The examiners want to be sure that you conducted this research and have a thorough understanding of it. Whilst the doctoral journey includes other individuals, such as your PhD supervisors, and peers, it is essential that you conduct the work and make the decisions. In the oral examination, as we will see in this book, the examiners are looking for you to be able to justify decisions made in your PhD and for you to 'own' your PhD. You as a student need to be confident and able to defend your work and articulate the decisions that were made in your PhD.

A further aspect of the viva exam being an oral examination is that the examiners want to be able to understand whether you can teach others about your PhD and its findings. Whilst the examiners may be in a similar or the same research area as you, they are not experts in your specific PhD research field and therefore in the oral examination you will need to be able to 'teach' them about your specific research. Whilst they may not need teaching about the methodologies, they will need to learn more about your specific PhD.

A fundamental component of the oral examination is for examiners to understand and explore the training that you have embarked on during your PhD. Conducting a PhD is a journey that allows you to explore a range of opportunities, and the competencies and skills that are developed in a PhD, regardless of the subject, are transferrable to any career that you choose to embark on post PhD, whether that be as an academic, a clinician or in another setting such as industry. In the written theses it is often sometimes not possible to fully explore and reflect on the training opportunities that the PhD enabled you have completed and so the viva is the opportunity for you to share these with the examiners and how they strengthened the PhD.

In a viva exam, examiners want to be able to hear the work you have conducted and what it adds to the research field. With many doctoral students being funded by government and the public such as through the UKRI, it is important that the work that has been conducted makes a contribution. Often the assumption is that, a doctoral student's work has to be groundbreaking; this is not the case, however, but it does have to be novel and original and to make a difference.

2.5   The Viva as a Dyadic Conversation Not an Interview

One of the key components of a doctoral viva is that it is a two-way conversation between you as the student and your examiners. This is an aspect which is similar across all doctoral vivas regardless of disciplines and HEIs. It is important to note that the questions asked to serve as discussion point as opposed to a requiring a question and answer response one may expect in a job interview setting. The examiners' questions are asked as a way of prompting discussion and so when considering these, as we will delve into it in greater depth throughout the book, they should be answered in a way that shows both independent and intellectual thinking. Within the viva, as a student there will opportunities for you to lead the discussion in a way that you wish to, and thus it is crucial that you are able to think 'on the spot' in an intellectual way.

It is important to be aware that the viva examination is often led by the external examiner, as they are independent from your HEI and organisation. The internal will, however, be able to ask questions throughout the examination.

2.6   The Viva as an Opportunity to Engage with Academics

Depending on your institution's policy concerning the conduct of doctoral vivas, the viva will be an opportunity to share your PhD findings with academics with whom you have not previously interacted. This is an opportunity to meet new academics and share your findings with academics outside of your direct research field perhaps too. Some doctoral candidates go on to complete post-doctoral roles with their examiners and thus the viva is an opportunity to make new connections that may help you in future career endeavours. For others, their doctoral viva examiners become collaborators in future research projects due to a shared interest. The viva is an opportunity to share your findings after all those years of conducting the research!

Having conducted your PhD over several years, and for some this has been completed part-time whilst having other jobs, The viva is an opportunity to share your findings after all those years spent recruiting participants, running analyses and juggling competing demands. It is an opportunity and experience that will never come again—which is a relief for some, but it should be enjoyed. These people will have read every word you have written and their attention for the duration of the viva will be focused solely on you and your work whilst this may appear daunting—value this experience.

2.7 Practical Points

In situations which are unknown and unpredictable, it is helpful to reflect on the factors that are known. These include, but are not limited to; choosing examiners, and the process post viva, such as how will you receive your results. It is important to note that with these practical points it is often the role of your supervisory team to organise these factors.

Choosing Examiners

This is a critical component to be aware of when approaching your doctoral viva. Often, this is something not approached until the final few months prior to submitting your thesis, but it is something that should be factored in as you approach your final year of the doctoral viva.

Often people do not know who their examiner should be, where to start and how to choose them. There is a common misconception that it is for the student to choose their examiners on their own; whilst this is partly true, it is not a decision that the student should make without their supervisory team. As will be discussed shortly, it is crucial that communication between you and your examiners is through your supervisor and/doctoral graduate college.

When choosing your examiners, think about; who you would like and who it would be interesting to share your thesis with, who may enjoy it and who is a researcher in a similar research field. Then, together with your supervisors, you can discuss who to invite. It is often advised to choose examiners with whom your supervisors are familiar. The supervisory team will approach the potential examiners, often through the doctoral college, and invite them to examine your thesis and doctoral viva. It may be the case that some cannot commit to being an examiner. This is not a reflection of you or your work; more often than not it will be due to competing demands on their time.

When choosing examiners, consider about whose papers you enjoyed reading in your doctoral journey, and whose insights may be interesting; but it is important to note that you do not have to have cited their work in your thesis. Many examiners do not work in the same specific field as students they examine, but this certainly does not make for any less of an interesting viva discussion.

Some HEIs have rules which mean that you cannot have worked with, or be involved with, the examiners in any way; similarly, some have rules that means that your examiner at your first-year transfer/continuation/upgrade viva cannot examine your final thesis and viva. Your own doctoral graduate school will guide you as will your supervisory team.

When Do my Examiners Receive my Thesis?

Your examiners receive your thesis once you have submitted this to your institution. They receive the thesis directly from your doctoral/graduate college; for some, this may be electronically in other cases a hard copy is posted to examiners.

What Do my Examiners Do Prior to my Viva?

Prior to your viva examination, your examiners will have independently read and commented on your thesis. Each examiner will then write a report independently and prior to your viva examination they may a have pre-meet to ensure that they are both aware of the other's thoughts. It is important to be aware that at this point the examiners will not have decided on your outcome.

When Should I Have my Viva?

After thesis submission, it is advised by some HEIs to have your viva examination within a set time period—for some this has to be within six months whilst for others it has to be a specific time period after thesis submission. It is generally advised to have your viva examination at least one month post thesis submission. This allows the examiners ample time to read and comment on your thesis and gives the opportunity to go through your work prior to the viva. It is important that the date is chosen and agreed prior to thesis submission if possible.

Timing of the Viva

One of the concerns that doctoral students often have concerning the viva is timing, with rumours that it could be over three hours or less than three hours. If it's less than three hours does that mean it's not good, or if it's over this time does that mean it needed more explanation? This uncertainty concerning the timing of the viva is not based on any factual information, such as the idea that a better thesis will require shorter time and vice versa. Sometimes the timing of the viva depends on practical issues such as: does someone have another work commitment or a train to catch after the viva? The timing cannot be stipulated prior to the viva due to nature of the viva; for example, the viva may lead to further questions and thus discussion may last longer. One thing to consider when asking others on their experiences of their doctoral vivas is to not be concerned about length of the vivas. At many HEIs, there are rules in place which mean that after a set amount of time there is a break for all involved in the vivas. You are not expected to sit for the whole duration without a break if you require a break.

Location of the Viva

Recently, many vivas have begun to take place online through video conferencing software. Vivas that take place online run in a similar format. Vivas that take place in person will usually occur in a neutral room on your HEI campus, rather than the office of either you or your supervisor. You will be notified of this room through your graduate school; it is not for you to arrange. It is often a good idea to familiarise yourself with this room prior to the viva.

Who Will Be in the Viva Room?

This varies from institution to institution. Often it is an external examiner, that is someone who is external to your institution or organisation, and an internal examiner who is someone from your HEI. Some institutions require the presence of an independent chair in the viva. The chair is an individual who is an academic at the PhD student's institution and their role is to chair the viva, but they are not there to enter into any discussion with the examiners and student and they do not play an active part in the viva. Some HEIs require an independent chair when one of the examiners have not previously examined a viva at that institution to ensure the process is standardised where possible.

Before you have your doctoral viva examination you will be informed that an independent chair will be present and this is something that you can clarify with your supervisory team and graduate school. The independent chair is selected by your supervisory team and graduate school and is not somebody that you should approach. You may also be given the option for your supervisor to be in the room. Whether this is an option for you may depend on your institution. For some, the presence of your supervisor might make you feel more confident; for others, it may not. It is worth noting, however, that, like the independent chair, your supervisor cannot take part in the discussion or help you in any way. In fact, if present, they are often seated out of your line of sight.

Taking Notes in the Viva

Examiners may make suggestions of changes or typographical changes in the viva, but these will be communicated to you post viva, and so in the viva itself your sole focus is on articulating and engaging in conversation with your examiners about your work. There is no need to make notes, and this is something that the examiners will do and communicate with you post viva. There is no expectation for you to make notes in the viva.

Communication Etiquette Regarding Viva Examination

Due to its unpredictable nature, the viva often results in candidates feeling bewildered by the process and so it is unclear who arranges what. The rules across all HEIs are that all communication between both examiners goes via your doctoral college/graduate school and your supervisory team. As a doctoral student, it is not appropriate for you to contact your examiners prior to the viva examination; your focus is on the viva rather than administrative details.

2.8   Post Viva

Often the focus is on the preparation of the viva examination and there is little information about what happens post viva. Unlike many other oral examinations or job interviews, in your doctoral viva, you will be given your result there and then. As a student, the examiners will ask you to leave the room after you have had a discussion on your thesis with them; they will then invite you back into the room and will tell you your outcome. They may explain corrections then and there; however, it is important to be aware that these will be communicated to you in a letter post viva from your graduate school. Again, at this point, all communication is through your doctoral school rather than directly with examiners.

Summary

- The viva is an oral examination.
- The viva is an opportunity for you to engage in a conversation with senior academics.
- Communication regarding the viva process is via your doctoral college/graduate school.

3

Urban Myths about the PhD Viva

Stacey Bedwell

3.1 Viva Rumours

Everyone has heard at least a couple of rumours when it comes to the PhD viva. It's natural for rumours to circulate when it comes to what is essentially quite a mysterious process. These rumours are often no different to the urban myths many of us heard in school, like the boy who put a frog in a blender or the child that microwaved a cat—these were the urban myths that circulated in my town; perhaps yours were a bit more animal-friendly! My point is, most of the stories we hear about how awful the viva is are just that, stories. Some common whisperings in PhD offices around the country include the ten-minute viva and the monster examiner. The release of *The Theory of Everything*, a movie depicting a very short viva exam, did not really help with this rumour. The reality is that a ten-minute viva is not impossible, and they have happened. In my own experience, however, more than ten minutes had already passed by the time the independent chair had gone through all his paperwork and explained the process, so it would have been a very impressive feat to be awarded a PhD after not saying more than a polite "hello" to my examiners. As for the *monster examiner* experience, which we hear of so often, this one is also technically possible, if your examiner is perhaps in an awful mood that

S. Bedwell, I. Butcher, *How to Excel in Your Doctoral Viva*,
https://doi.org/10.1007/978-3-031-10172-4_3

day or you have a personality clash—what we each interpret as a monster is, however, quite subjective.

In the following sections I have highlighted some of the most common *viva myths* and debunked those that really are just myths aimed to scare us.

3.2   Top 20 Urban Myths about the PhD Viva

As you read through the following twenty common viva myths, see if you can identify which ones are true and which are false.

1. *The ten-minute viva*
   The PhD candidate who was only in the exam room for ten minutes before being awarded their PhD.

2. *The ten-hour viva*
   The PhD candidate whose viva exam went on into the night, the examiners asking about every minute detail of the thesis.

3. *The monster examiner*
   The examiner who seemed to want the student to fail.

4. *The pass with no corrections*
   The student who wrote a thesis so perfect, they didn't have to change a thing.

5. *The failure*
   The student that did not pass the viva and failed the PhD.

6. *The MPhil*
   The student that was awarded an MPhil rather than a PhD.

7. *The student that got a job from their viva*
   The PhD candidate that was offered a postdoctoral position working with their examiner.

8. *The student that broke down in the viva*
   The student who experienced an extreme emotional reaction to a subject of discussion in the viva exam.

9. *The impossible question*
   The examiners that ask the candidate quiz-like questions about irrelevant or very far-removed information that is not related to the topic of the thesis. An alternative version of this myth is where the examiner asks the candidate to mentally calculate a very complex formula or statistic.

10. *The quiz-viva*
    The examiners quiz your scientific knowledge, that they expect you to know everything about the scientific discipline you study in.

11. *The bored examiner*
    The examiners become visibly bored whilst discussing your research.

12. *The smooth talker*
    The candidate who could 'talk the talk'. They managed to talk the examiners round to a different outlook, opinion, or viva outcome.

13. *The defensive viva*
    The PhD candidate who took every comment personally and struggled to take on board any critical feedback.

14. *The enjoyable viva*
    The PhD candidate who recalls their viva as one of the most enjoyable experiences of their life.

15. *The trick questions*
    The questions designed by the examiner to catch the PhD candidate out, or to trick them into saying something incorrect.

16. *The lapse in memory*
    The student that forgot absolutely everything they had done throughout their PhD upon entering the exam room.

17. *The wrong answer*
    The examiner asks a technical question, a question about previous research or any other specific question and the student gives an incorrect answer.

18. *The viva that turned it around*
    The case where the student performed so well in the viva they changed the examiner's mind. This can also go the other way—the student who performed so badly in the viva they changed the examiner's mind.

19. *The disagreement*
    The student that started an argument with the examiners when they disagreed with an approach, method or interpretation.

20. *The full year of corrections*
    The PhD candidate who spent a year on corrections after their viva.

3.3   Debunking Common Viva Myths

Now we have identified some of the common myths that circulate postgraduate offices and groups, I'm going to offer a little explanation for each of them. Some of these myths are simply that, myths, but some that we might assume to be untrue or impossible do happen. One of the best things you can do to prepare for your viva is to educate yourself as to what might happen and what is very unlikely.

1. The ten-minute viva

False!

Technically speaking, your viva exam could be any length of time, and I have heard of some very short ones (note that I was not part of them and did not witness them, so cannot guarantee these are true accounts). So, theoretically you could be in and out of the viva in as little as ten or twenty minutes. However, this is very unlikely. Personally, I would feel quite insulted if my examiners didn't want to talk about my work for a bit longer; even one question on each chapter is going to take longer than ten minutes, probably closer to an hour. It's also worth considering the general housekeeping and administration you need to go through at the beginning and/or end of the viva itself. For example, in my viva the independent chair took a good ten minutes explaining the general process, what was going to happen, what the outcomes could be, and so on. If my exam had ended after ten minutes, I wouldn't have said more than a polite hello to my examiners. Consider that your examiners have spent a good deal of time reading your thesis; for me, this was nine chapters and eighteen appendices. If they have an interest in your area of research (which they should do), they are going to want to learn about what you did in each of your studies in a little more depth than can be written and learn more about how you think your work fits into the bigger picture. This simply cannot be done well in a very short space of time. So, whilst a very short viva is theoretically possible, it's unlikely to happen and you don't want it to.

2. The ten-hour viva

False!

As with the ten-minute viva, a very long viva exam is technically possible. However, your examiners know you are only human, as are they. Nobody wants to sit in a room engaged in intense conversation for more than a few hours. It is my opinion that nothing can be achieved in an excessively long viva that could not be achieved in an average three-hour one. There are only so many questions that can be asked, and once your examiners have established the answers to the main aspects of your work they aimed to discuss with you, there is no need to continue; this would not add any value and it would not change the outcome of your viva exam. It's also worth reminding yourself that your examiners also have lives to get back to; often they have travelled across country for your viva and have a train to catch. Logistically, taking longer than a few hours would be inconvenient.

3. The monster examiner

False!

It's all too easy to assume your examiners are there to catch you out and to tell you how awful your research is. The truth is, there is really no reason why your examiners would benefit from being awful to you and about your work. It's understandable where this misconception comes from, especially if you have been through the process of blind peer review with an academic journal—most of us have heard of the terrible reviewer 2. Like many academics, I also have had some pretty negative experiences with reviewer comments from the peer review process. Some comments from reviewers can feel very much like personal attacks; actually, sometimes they *are* personal attacks. The book is not the place for me to get into everything that is wrong with current processes in academic publishing, but I can highlight that in most cases the review process experienced through a PhD viva is much more pleasant than the experiences many of us are all too used to when it comes to journal peer review.

For the most part, I think the difference in experiences between peer review and viva comes down to anonymity. Most reviewers for journal articles are anonymous, in the sense that you don't know who the infamous reviewer is until final publication, if at all. The authors are usually anonymised for the review process as well. This practice means the reviewers are very distanced from you as a human being. This is not the case in your PhD viva—your examiners are literally sat in front of you whilst discussing your work. It goes against a lot of natural human behaviour to attack you and your work in this situation, so it's very unlikely to happen—even if you feel like you have been attacked by reviewers in the past.

4. The pass with no corrections

True!

This one does happen. As PhD students, most of us tend to be told early on that a *pass with no corrections* is not possible. After all, there will always be at least a typo. When it comes down to making outcome decisions, it's important to be aware that there is no set marking criteria or grading scheme for the examiners to follow. They do not sit there counting the errors and typos to decide which outcome to give you. For this reason, each outcome is very individualised. In the case of no corrections I mentioned, there were typos in the thesis and probably a couple of things that could have been rephrased more to the examiner's preference. However, in this case the examiners deemed that the small number of changes they would ask for were not worth going through the process of minor corrections. There was nothing wrong with the thesis and the message it conveyed; this is the important factor.

5. The failure

True!

As an academic community, we often don't talk about the more negative outcomes of the viva. Similarly, in the same community we rarely shout about the job interviews that went badly, the grants we didn't get or the papers that got rejected. Perhaps this is something that we all need to strive to do more of, to give a more accurate representation of life in academia. Disappointing viva outcomes fall into the same category of things that happen in academia on a regular basis, but we don't send mass department emails about, we don't tweet about, and we don't tell all our friends about. This does not mean it never happens.

Failing the viva is a real possibility and we should never assume it cannot happen. For most PhD candidates, your supervisory team will not intentionally send you into a viva if they are not confident you will pass. However, sometimes even the most confident candidate can have a bad experience in the viva. In most UK institutions and doctoral colleges, there are procedures in place that intend to prevent this from happening. For example, you have probably gone through some form of transfer from MPhil to PhD or an upgrade viva in your second year of full-time study. In this phase you demonstrated that your work was that of PhD standard to an internal committee or review panel. This does not guarantee, however, that your viva examiners will agree. There is always going to be this element of unknown in the PhD viva.

It is important for me to clarify that although a failed viva is possible, it's very rarely the end of the road. Usually if your examiners are unable to award you a PhD, you will be invited to resubmit your thesis and sit the viva exam again. This is known as fail and re-sit. You would receive corrections to be completed, just as you would have done if you had received minor or major corrections, but they will likely be more extensive. Your thesis will be resubmitted, and you will go through the viva again. If you satisfy the examiners, they will go on to award you your PhD. You still might receive minor or major corrections after the second viva.

6. The MPhil.

True!
Much like failing, or re-sitting, being awarded an MPhil. after your viva is a real possibility. This is another outcome that is not often talked about, because people don't tend to brag or post on social media about their MPhil. outcome. It probably happens more often than you think. If you have a supportive supervisory team and a good support system in place at your institution, e.g. your doctoral college, you should never be in the position where you are entering your PhD viva and you are not at PhD level. This doesn't mean it can't happen though. Most UK universities now follow a clear structure to the PhD. You probably completed a proposal or project approval within the first months of your studies. You then likely went through several reviews or reports at set increments, e.g. every six or twelve months. Many students also go through a transfer or

upgrade process, where you actually start off on an MPhil. and transfer to a PhD in your second year. At this stage if you completed it, you demonstrated to a review panel that your work was PhD standard and beyond that of an MPhil. With this in mind, if you have already demonstrated your research is PhD standard, there's no reason why it should be deemed lower than PhD standard when it comes to your viva. However, we shouldn't be over-confident; things can go wrong, and your external examiner could review your work differently to internal panels. I recommend in preparing for your viva you ensure you have a clear answer to "What have you done that deserves a PhD?"—there is more on questions you should be able to answer in later chapters.

7. The student that got a job from their viva

   True!
   This is a myth that was thrown around a lot when I was a PhD student. Personally, I have not spoken a word to my external examiner since my viva and the only communication I had with my internal examiner was regarding my thesis corrections. This doesn't mean you won't ever end up working with your examiners, however; my experience is not necessarily yours. For me, neither examiner's expertise was 100% aligned with my research, so working with them would have been a bit of a stretch. The case is very different for others; I know of at least one person who went to work with their external examiner on a postdoctoral project very soon after their viva. If you think about it, it makes sense in a lot of cases. If your examiner is working in the same specific field, and they happen to have a postdoc position available as you come to the end of your PhD, you would likely be applying for that position, whether you knew the PI as your examiner or not.

8. The student that broke down in the viva

   True & False!
   Everyone cries at some point in their PhD; it's a rite of passage. For me, it happened in my first year. I literally ran out of my supervisor's office. That rite of passage has probably been and gone well before your viva. At this point, you've got all the data, you've done the analysis and you've writ-

ten about 80,000 words about one tiny thing. The viva is the celebration of all that, not the time to have an extreme emotional response. The conversation with your examiners is unlikely to make you react this way. Afterwards is a different story; coming out of that room at the end of four years of such intense work is a huge emotional relief, and that can be overwhelming!

Just because I've never heard of it happening, does not mean it does not. Nobody can really predict what might be said or what might trigger you to react in an emotional way, especially if you have other things going on in your life—as much as it may seem so, nobody does their PhD in a bubble. Life continues around us. If you are the kind of person who has emotional reactions in supervisory meetings or if you have a tendency to take criticism personally, you might benefit from considering an emotional reaction in your viva as a possibility. If you are prepared, it's not a surprise and you will be able to deal with it in the moment.

9. The impossible questions

False!

I am guilty of believing this one. I was convinced entering my viva that the examiners were going to be asking me very difficult questions about specific biological processes, complex analyses and the details of every research study I had cited. I was also convinced I needed to know all of this by memory. Now, I knew I would have my thesis there to look at and I knew I would also have the notes I had made for myself in the margins. I *knew* it wasn't going to be a test of memory. My anxiety got the better of me, however, and all reason went out the window. I convinced myself my viva would be different, and that my examiners would be disappointed if I needed to check something in my thesis or I didn't know some obscure detail by memory. Logically, this was never going to happen. Examiners have no interest in asking you questions there is no reason for you to immediately know the answer to. The examiners know how the viva works and they are not there to test your ability to memorise obscure information; they are there to establish firstly that you did the work, and then that you are able to explain and discuss the work with others at an intellectual level. If you are uncertain of the answer to a question posed by the examiners, it's fine; this won't make or break the outcome. If you can't answer anything, the examiners ask you, of course, that is a different story. But remember, you've been doing this research for three or four years. You wrote the thesis. You know it.

10. The quiz-viva

False!

It's quite a common misconception that the viva exam is a test of your knowledge. I fell for this one myself. If you recall from chap. 1, I took on a PhD project in neuroanatomy, having not studied biology in any great detail since A-level. I don't really like using the term imposter syndrome, but I had a feeling that I didn't know enough biology. This feeling stayed with me throughout my PhD, including during my viva. It wasn't helped by a meeting in my first couple of months when an internal reviewer pretty much confirmed my fears that I didn't have the skill to be there (that was a lesson in how not to support students). As a result, I overcompensated when it came to preparing for my viva. I could recite very detailed biochemical processes that I really did not need to know. I learnt all this because I was convinced the examiners were going to quiz me on how specific reagents I had used worked in minute detail, far beyond what would be sensible for me to know as an anatomist. I was a little disappointed when nobody asked me!

My point is, it's not a quiz. The examiners are not there to quiz your knowledge in a test of memory. They are there to have a conversation about the research you did. The reality of the situation is that you probably don't need to memorise a thing; you've been doing this project for several years, and nobody knows what you did better than you. If there is anything within your project that's a little complex and you can't quite remember, your thesis is there in front of you to check what you wrote.

11. The bored examiner

False!

As a PhD researcher you are more than likely very used to talking about your research with people who don't understand it, or perhaps I should say talking at people who don't understand it. In my own experience lay people tend to show signs of boredom as soon as I start talking about systematically measuring the three-dimensional location of thousands of individual labelled neurons within the cortex. I'm not surprised; it doesn't sound that thrilling if you're not an expert in neuroanatomy. These experiences, which you may have had quite a lot of in the lead-up to your viva (we do consistently tell you to talk about your research with anyone who will listen), can give you a false impression that your work, and the way you explain it, is dull. It's therefore understandable that you might think the examiners are going to get bored and zone out as you are talking about your research in depth for maybe three or four hours. Remind yourself, though, of who your examiners are; unlike your brother Dave the lawyer or Auntie Nora the novelist, both of your examiners are research scientists working in an area either in or very close to the specific topic of your thesis. This means they have a genuine interest in what you have to say. If your examiners are working in your specific area, your findings may link very closely to what they are doing and may inform their current projects, so they are quite unlikely to get bored.

Getting tired, on the other hand, is a different matter. No matter how invested we are in a given topic, talking about it for three or four hours non-stop is tiring. For this reason, it's important you remember you can take a break, and your examiners can too.

12. The smooth talker

True & false!

Whilst it's not inaccurate to say the viva can make or break your PhD, it's not going to be possible to talk the examiners into a pass with no corrections if you reported an analysis incorrectly or missed out a vital reference. The way you discuss your work, and the extent to which you are able to demonstrate your ability to think, can play a big role in the impression your examiners are given overall of your PhD and, ultimately, whether a PhD is awarded. This can go both ways, but it is worth bearing in mind that most examiners begin your viva already knowing what the outcome is.

13. The defensive viva

True!

Defending your thesis and your research is a vital part of the viva. However, there is a fine line between defending and becoming defensive. It is possible to take a comment or question personally, especially when this is four years of your life being discussed; getting defensive about something we are quite attached to comes naturally. Considering my own PhD viva as an example, there is one particular point that stands out as a time I almost became too defensive about my research decisions. I overcame this by following up my natural defensive reply with a clear justification, explaining why I thought my decision was correct. If I was not able to do this, I likely would have given my examiners the impression that I could not consider other perspectives or criticism—important attributes to an independent researcher. Being able to avoid becoming defensive with your examiners is important if you are to demonstrate your ability to think independently.

14. The enjoyable viva

True!
It might sound impossible when you are preparing for your viva and you are built up with anxiety, but it is possible to enjoy your viva. We'll cover steps you can take to ensure your viva is an enjoyable experience in later chapters of this book. It can be hard to step back from the feeling of it being an exam, and the pressure PhD candidates can often feel on the culmination of several years' work, but if you can step back you can appreciate the unique nature of the situation and appreciate the opportunity in the moment.

Not everyone will enjoy their viva, and that's okay. Many of you may enjoy the feeling of passing the viva much more, or even feel a sense of relief. For some, the viva exam itself is enjoyable because it is an opportunity to talk about an area of research you are passionate about for an extended period, with the undivided attention of experts in the field. It's a great time to debate current controversies in the field, to receive input on your research for future developments and to consider new insights.

15. The trick questions

False!
This topic will come up again and again throughout this book, and for good reason. Many students head into their PhD viva expecting the examiners to try and trip them up. Whilst we can't promise it won't happen, we can't guarantee what an individual will and won't do on the day, we can say this is never the purpose of your viva. For the most part, all viva exams tend to follow the same tried-and-tested format. The external

examiner usually leads the discussion, following a systematic approach. In most cases, the examiner will work through your thesis from beginning to end, asking questions about each chapter as they go. The points they want to ask you about will likely be decided well before you sit down in the actual viva—you might see sticky notes in the printed thesis to indicate where these questions will be. Questions will often have the intention of the examiner being able to gain a more in-depth understanding of your research. As we've mentioned before, they are not intended to be quiz-style questions. Everything you are questioned on you have done, it's your research, you know it better than either of your examiners, so if anyone would be able to come up with a trick question, it would be you!

16. The lapse in memory

True!

A lapse in memory when it comes to a specific detail, or even whole studies, is possible and does happen. It is not a bad thing, however. Not being able to recall something on the spot when you are under immense pressure and probably feeling quite anxious should never reflect badly on you. Your examiners understand that it's a tense situation for you. If you can't recall something in the moment, it's fine; that's why you have your thesis in front of you. Do yourself a favour and follow along with the examiners. Chances are, your examiners are going to work through your thesis in the viva in a systematic way, starting at chapter 1. When the examiner turns to page 137, you turn to page 137. This takes the pressure off when it comes to recalling the specific detail the examiner might be referring to on the page. If you need a minute or two to read the page back, that's OK. You're talking about several years' worth of work and a few hundred pages here; nobody expects you to have memorised the whole thing.

There are times when a lapse in memory could become problematic; if you struggle to remember the basics of what you did, why you did it and what the value of your contribution is, you might fall into trouble.

17. The wrong answer

True & False!

Firstly, it's important to remember that the viva exam is not a quiz with a mark sheet on which the examiner is adding up points. They will not be sat across from you counting how many things you get wrong. The viva really is a conversation between experts, and a conversation about a developing area to which you are making a new construction at that. When it comes to interpretations and conclusions, there is not really a wrong answer. You might differ in opinion, and that is fine. If you interpret something differently to the way your examiner does, neither of you are wrong—this is where you can have a healthy debate. It is important that you are able to show your ability to think independently and to consider multiple perspectives, so you will do well to discuss why your examiner thinks your interpretation is wrong and justify why you think you are right. It's also OK if you change your mind. Sometimes with hindsight and increased knowledge, our understanding of findings can develop. This is how science works after all! If you agree with the examiner's point, explain why and ensure you explain why you changed your mind/why your original approach was decided upon to begin with.

There are other instances in which your answer could be wrong. For example, perhaps you mistakenly applied an incorrect statistical test or perhaps you mis-remember something and give the wrong information. This is OK too. Your examiner knows you are human and doesn't expect perfection. It is important that you can take ownership of mistakes, whether they are in the thesis itself or simply something you said in the viva. If the examiner points out something that is incorrect, and they are right (or even if you are unsure), take that on board and acknowledge what it means. You have a chance to make corrections after your viva; as long as it isn't a major flaw in your whole project, it won't be a big deal. Many corrections examiners ask for are made to improve the quality of your thesis and subsequent papers.

18. The viva that turned it around

True!

Have you ever considered why you are doing a viva exam at all? One reason is to demonstrate that it's really you that wrote the thesis. It's also necessary to show that you can discuss your work and its implications. If you wrote a fantastic thesis but could not articulate concepts well or struggled to show evidence of thinking independently in the viva, this could lower the examiners' opinion of your contribution. Similarly, you could write an underwhelming thesis and present yourself and your work very well in the viva, thereby improving the outcome. Sometimes, examiners might not get a complete understanding of the relevance of your findings in a written format, but you may be able to explain and answer their questions more fully verbally. For these reasons, the viva and the conversations you have in it can turn the outcome around—despite the fact that most examiners go into the viva knowing what the outcome is likely to be. You might surprise them.

19. The disagreement

True!

It can be easy to assume that you need to agree with everything your examiner says to do well in your viva exam. However, I would argue quite the opposite. Blindly agreeing with the examiners could give the impression that you are unable to think independently, or, as we say so often, to defend your work. I've had conversations with many PhD candidates worried about disagreeing with their examiners, often thinking that if they don't agree they will fail. It's hard to remember sometimes that you are the expert in the room when it comes to your PhD viva. Your examiners might be experts in the field, they might even be world-renowned researchers with a few *Nature* publications under their belt, but they still didn't do your project. You did. Part of the viva is demonstrating your ability as an independent researcher, and an important factor in becoming an independent researcher is being able to think for yourself and to have an intellectual conversation.

The disagreement happened in my viva. One of my examiners questioned a specific statistical test I had applied. I was confident in the choice I had made and explained why—perhaps quite passionately, but scientifically, not subjectively. As it turns out, the examiner had misinterpreted something important about my data and this is why he thought I had applied the wrong test. This resulted in a minor correction in how I explained my data, to improve clarity. So, this disagreement (or, more accurately, intellectual conversation) was important.

20. The full year of corrections

True!

This one can and does happen. The length of time you are given to complete corrections is not set in stone. There may be some general guidance set out by your university; for me minor corrections were generally given three months and major corrections six months. This can vary greatly, because each thesis and the corrections needed are individual, as are your personal circumstances. It's worth bearing in mind that the length of time you have for corrections can be put down, at least in part, to a conversation between you and your examiners. For instance, whether you need three or six months could come down to how much spare time you have in the near future. If your examiners want you to collect more data, and the nature of your research means data collection is very time-consuming, you might need longer than six months. Similarly, if you need to rewrite large sections of your thesis and you are working full time in a new job, this is going to take a long time. Sometimes, if the decision at the end of your viva is to resubmit the thesis and complete another viva, this might take up to a year.

Summary

1. The viva process, although different for everyone, does follow rules and regulations.
2. Examiners are human and there is no benefit to them being unreasonable.
3. For the most part your examiners are there because they are interested.

4

Real Viva Experiences

Stacey Bedwell and Isabelle Butcher with contributing
authors

In this chapter, we invite you to meet PhD graduates from a wide range of disciplines, institutions, and geographical locations. You will learn about the real experiences, some positive and some negative, from short vivas to very long ones, and a wide range of outcomes. Some of the individuals who share their experience with you completed their PhD viva recently, others much longer ago. Also, within this chapter, you will learn about the viva experience from the perspective of the examiner—a unique and valuable insight to have as you prepare for your own viva.

© The Author(s), under exclusive license to Springer Nature Switzerland AG 2022
S. Bedwell, I. Butcher, *How to Excel in Your Doctoral Viva*,
https://doi.org/10.1007/978-3-031-10172-4_4

4.1   General Viva Feelings

Giulia Furini

"I completed my PhD at Nottingham Trent University. Before the viva, I was a nervous wreck. I might have cried a little before the start. It all went away immediately after it started and I really enjoyed the experience. In the viva itself, there were the two examiners, which reviewed my thesis and asked the questions, and an external member, which witnessed the whole exam and was needed to validate the legitimacy of the exam. I do not remember exactly, but I would say my viva was somewhere between two and three hours. We did not take breaks, as far as I remember."

Damien Neadle

"I completed my PhD at the University of Birmingham. Prior to the viva, I was a wreck. I couldn't eat or drink. I have only ever been more nervous than that once—when my daughter was being born. However, after the viva the relief was indescribable. I felt as though I had time again, previously all the time was spent preparing and worrying about the viva. After, I took a week to do very little and reset (apart from working in my job as a lecturer). In the viva were my external examiner, my internal examiner, and the chair. It took around 3.5 h. We did take a short comfort break at the suggestion of the external examiner about halfway through."

Nicole Pearcy

"I completed my PhD at Nottingham Trent University. I remember being extremely nervous before going into the viva—I didn't think I'd be able to get through it but as soon as the viva started the nerves disappeared. I remember thinking it was much more relaxed than I expected but I was still so relieved, proud, and excited to have completed it."

Theodore Hughes-Riley

"I carried out my PhD at the University of Nottingham. I was nervous going into the viva and I felt quite overwhelmed afterwards. I think that I expected to leave feeling validated, and that the work I had carried out over the three years was worthwhile and done well. Minor corrections was really the best outcome that I could have hoped for (no corrections is pretty much unheard of in the sciences) so I should have been happy: I think that due to the build-up of stress leading up to the viva itself, after it was finally over I felt quite bad (instead of relieved as I had expected)."

Richard Fereday

"I completed my PhD at Cardiff University. I felt a little nervous before going into the viva, but not overly so. I didn't particularly like dressing in a formal suit, so this might have increased a sense of unease. Afterwards, it was a little relief, but not a huge wash of emotions.

Present in the viva I had an internal, the external, and the chair. The viva was around 3 h if I recall correctly, possibly just under. It didn't feel this long though—time went quickly. I did have a small break of around 10 min. I didn't leave the room though, didn't feel the need."

4.2   Subject-Specific Viva Experiences

Biosciences

Giulia Furini completed her *PhD in Biosciences* between 2013 and 2017.

"My project focused on the role of a specific enzyme, called Transglutaminase 2, in the context of chronic kidney disease and kidney fibrosis. Specifically, I was studying the secretion mechanism of this enzyme from the kidney cells to the extracellular space during the disease, using cellular models as well as a proteomic approach, to determine the enzyme interactions during its release.

It took almost four years to complete my PhD. To be specific, it took almost three and a half years to complete the experimental part, then another few months to write and submit the thesis. I submitted on the very deadline, because I was keeping on making little tweaks and corrections, I was never fully satisfied!

To prepare for my viva exam, I went through the topics in my thesis introduction and discussion, and I made sure I knew as much as possible of the subjects related to my research. For example, if my experiments were involving cell death, I was making sure I remembered the basic pathways of cell death and at least the main molecules involved, even if I my experiments were not specifically regarding those molecules. Honestly, I ended up studying much more than I actually needed for the viva, but it helped me feeling confident about my knowledge and my experimental work. Moreover, I tried to keep an eye on the examiners' background, since it is likely that your data might ignite some ideas in their head while they are reviewing your thesis, as they might find some connections with their work. It is good to play ahead and find connections too, as these subjects might emerge during the viva.

As part of my preparation, I had a mock viva with my head of the studies, which took place approximately a week before the real viva. I felt it was useful, but sometimes one can end up worried after the mock viva, given that your PhD supervisor might turn out to be much more demanding than your actual viva examiners.

I was not really surprised by any of the questions. I made sure I was quite prepared on most of the subjects related to my PhD project and the data discussion. Moreover, by trying to figure out potential connections with the examiners' field of study, I tried to be prepared to potential questions related to their specific background. I was pleasantly surprised, however, how most of the questions actually turned into a stimulating discussion with the examiners, which I really appreciated. Overall, I was surprised by how much I enjoyed my viva, compared to less positive expectations. You always think the examiners are going to bombard you with questions and make your life hard, but, in reality, it almost never happens!

Surprisingly, the viva was very pleasant. It was good to have an occasion to discuss your work, its pitfalls, the troubleshooting, its connections to other subjects, in a stimulating environment where the discussion can progress in length and is challenged by the examiners' questions and suggestions. Well, they had to stop me few times because I was talking too much, but that was fun too.

After the viva, I remember I was happy and relaxed (cannot say satisfied, as I actually never am fully), however crazy tired with the adrenaline dropping.

I passed with minor corrections. I got them on the day of the viva itself and I completed them very quickly, probably within a week or two, as they were only minor typos and a few parts to clarify better. To be honest, I do not remember much of the corrections. I remember they were mostly typos and grammar corrections. A few parts of the discussion, also, needed additional explanation. Yet, they were minimal modifications that I could complete rapidly. I remember there was a word that I misused throughout the whole thesis (dozens of times!) because it sounded like the literal Italian translation, my native language, but had a complete different meaning! Apparently, I used that word incorrectly consistently throughout my whole PhD life and never noticed I was wrong!"

Ekta is an experienced researcher and currently works as a Partnership Manager in drug discovery. Ekta completed her PhD in 2017. Ekta shares her experiences.

"I undertook my PhD at Sheffield Hallam University at the *Biomedical Sciences* Research Centre. My PhD was funded by and in collaboration with GlaxoSmithKline (GSK). I developed methods to detect peptide/protein biomarkers from fingermarks via a shotgun proteomic approach involving in situ proteolysis and mass spectrometry. Further work was carried out for the implementation of bottom-up strategies into conventional workflows with the analysis of blood-specific proteins. Alongside my primary research area, I conducted research for the respiratory team in GSK; in order to address the need for non-invasive techniques to monitor patient compliance, I also further developed methods to detect drugs and peptide biomarkers from the sweat in fingermarks. My PhD was a four-year BBSRC case studentship funded by

GSK. I spent 3 years 8 months in the lab and wrote my thesis in just over three months. My thesis was submitted by publication, as I had published three first author primary research articles.

To prepare for my viva I read my papers and thesis numerous times to try and identify areas that may spark discussion. I read about the analytical techniques I used and the general area. I also read some papers written by my external examiners group. I did not, however, have a mock viva, as my PI was on maternity leave towards the last five months of my PhD. However, I had an informal call with two people from my group who had completed their viva a few years before. They advised me on areas to revise and gave me an idea of the questions that could come up. I was informed that the first question may be 'Could you describe your key findings in 10 min?' or 'what are the highlights of your PhD?' Prior to my viva, I was incredibly nervous; even though I knew I had spent time revising, I felt underprepared.

In my viva, an external and internal examiner and an invigilator were present. The viva was just under four hours, and we did not stop for breaks. I was surprised by some of the questions as I did not expect to spend so much time on the fundamentals and theoretical aspects of my technique. I was expecting there to be more focus on my work and findings.

The viva was not the most enjoyable. In hindsight I wish I had provided my PI with the names of a few potential examiners (that I met at conferences etc.). This could have made a difference to the whole experience. When disseminating work at conferences via posters or oral presentations you tend to meet so many incredible people. It is a great idea to connect with them for moments like this!

I passed with minor corrections and received these via email and completed these within a month or two. My corrections included making table legends more descriptive and the paragraph structure needed to be amended. There were also a few typos. After the viva finished, I felt relieved. I was just glad I got through it. I feel that I could have benefited from a mock viva. It was just unfortunate that I did not.

My positive viva memory was that I had published three peer-reviewed first author papers during my PhD and a literature review. Acknowledgment of my dissemination efforts was a good feeling."

Animal Behaviour

Damien Neadle completed his PhD between September 2016 and September 2019.

"My *PhD was in Comparative Psychology and Cultural Evolution*, meaning it spanned psychology and biology (specifically animal behaviour). So, I had supervisors across the two schools. My PhD took three years, I handed in on my very first day of the minimum registration period as I had a job lined up.

My PhD consisted of five empirical studies and a literature review. Each of the studies examined a different area of the question 'What drives the evolution of culture in animals?' I initially examined the notion of culture: 'What kind of culture can we assume that animals have?' Then, I considered the idea of cultural universals and the idea that phylogenetically similar species (e.g., great apes) might share cultural traits, stemming from their last common ancestors. I then tested the theoretical concepts identified through the previous two chapters, by designing a novel stepwise experiment, to examine the learning processes behind a supposed 'peak' of culture. I then designed an experiment to consider why the results of the previous chapter might be so; this involved a large-scale multi-species study at the Max Planck Institute for Evolutionary Anthropology in Leipzig, Germany. Finally, I pulled together the various factors identified using a naturalistic observational study of a semi-captive group.

To prepare for my viva I went through my entire thesis and wrote possible questions with answers on Post-it notes. I did this page by page. Any time I identified a key problem/issue with the argument, I would look back to the literature to consider the other possibilities and prepare to defend the choices made. I also spoke to friends and colleagues about what experiences they had during their viva and 'common' questions. I then prepared set answers to these questions. Though, these did not come up for the most part. Equally, my 'issues and possible questions were rarely considered and often the issues the examiners identified were not ones that I had considered.

I had a mock viva to an extent; it was not a viva with subject experts as my supervisor had left the university at that point. However, it was a mock with very experienced academics in psychology, who were able to give me a good idea of the format etc. so that I wasn't surprised.

I was surprised by some of the questions. There were ideas that I had not considered and theoretical positions about some of the models that I included that surprised me. There was one comment from the external examiner that shocked me in how frank it was.

On the whole, I can say I did enjoy my viva. I found it an interesting experience that has left me feeling prepared for virtually any job interview. The process leading up to the viva made it seem worse than it was. I was very nervous going in as I had built it up into more than it was. My examiners were both very kind and set me at ease, despite sometimes asking some challenging questions. I enjoyed the opportunity to talk about my research with experts who were not necessarily from my lab (thus bringing a new perspective).

I received an outcome of minor modifications. I received my corrections via email. There was a document that compiled the list of corrections from both examiners, this was sent to me via the school administrators. I completed these within the month allocated to me. Some of my corrections included reframing a chapter, minor typographical errors, inclusion of studies that had been overlooked, reconsideration of definitions in the light of some of the newly included literature and more visualisations added to one chapter."

Psychology

Richard Fereday began his *PhD in* P*sychology* at Cardiff University in 2012.

"I submitted in 2016 and finished officially in early 2017. My thesis was in the realms of cognition and perception, focusing on time perception and causality. Specifically, I investigated temporal (causal) binding, which refers to the temporal attraction of causally related events. This effect might be rooted in event perception, such that the causal trigger is perceptually shifted *forward* in time, or the outcome perceptually shifted *backward* in time (or both). Alternatively, it might be due to a contraction of time between causally related events. I investigated the latter approach and found a temporal contraction of perceived time between cause and effect. The experiments took three years to conduct, with a final year for writing up.

I did not do a great deal really to prepare for the viva. The project was fresh in my head, and I had already published my first article. This wasn't easy but dealing with the many rounds of revisions really prepared me well. I also had a low-key mock viva. My viva was in the early January, but I remember that it wasn't dominating my thoughts too much over the Christmas break.

I did have a little mock viva, nothing too formal. As said above, the publication process quietly added to my confidence and certainly helped my experience of defending research.

I was not surprised by any of the questions. There was one question that kept being pressed—no matter how I explained, it wasn't satisfactory. But I got there in the end. As said, the publication experience took most of the surprise away.

In a way, I think I did enjoy my viva. It didn't feel stressful, more like a chat about a common interest. To be honest, some of it could even have been done in the pub. I think a viva should feel like this. You can assess a candidate without inviting an atmosphere of fear.

My outcome was minor corrections. They considered giving me a week to complete the corrections, such as they were, but gave me the standard three months. I received my corrections via email, just under a week following the viva. I think I took only a week to complete the corrections. I could really have completed them with a good day's work, but I spread them over a casual week. Most required a sentence or two to clarify points. There were no lengthy edits (such as subsections or paragraphs to rewrite). I had a comment about improving the conclusions to subsections to better tie them together, and to better signpost readers. Another comment asked for a better description of a cited study's method. Others include better describing a Just Noticeable Difference, including support for an assertion, and clarify how a model's processes explain an effect. A large amount of corrections were trivial points such as missing punctuation marks and abbreviations on figures."

Elizabeth Lewis is a trainee clinical psychologist in central England and completed her PhD between 2013 and September 2016 (with her viva in Jan 2017) based in *Business and Management* at a university in the north-west of the UK:

"My PhD was looking particularly at the management of social care for older adults. I explored how management practices (Human Resource Management and Human Resource Development) could be improved to support care workers. I completed my PhD in 3 years and 1 month.

To prepare for my viva, I reread my thesis. Made prompt cards from expected questions found online and practiced my responses with my partner. I also marked each specific section of my thesis and important areas, such as rationale for studies, philosophical underpinning and design used, etc. Mock viva (please see below). We also had a kind of mock viva in the first year where we had to defend our research idea around the time we submitted to ethics. This involved a panel of staff from the department that we knew.

I had a mock viva about a month before my real viva. This was run by someone in the department who had recently been through his viva, with my main supervisor observing. Both gave feedback at the end.

I will never forget walking in and being offered a buffet of food in the room where the examiners had set up and just not being able to consider eating! Lots of nerves and adrenaline, I guess! I felt fine in the viva and then anxious during the deliberation.

All three of my supervisors sat behind me in my viva! One chair and two examiners (one internal, one external). While it was nice having my supervisors there, there were definitely times where I thought I might have preferred just my main supervisor there. It felt a bit pressurised at first, but I soon forgot they were present. My viva was two hours. We had a break only for deliberations which was approx. 40 min. Then we were called back into the room.

I think I was surprised that I was able to answer the questions, as I had been worried that I might 'freeze' and forget everything I had written. I was also surprised that there weren't more questions about general HRM. I found that the questions were very reasonable and not at all anxiety-provoking.

I did enjoy the viva, yes! Although it was an unusual kind of enjoyment; I felt very prepared and ready for the viva but there was always that doubt that I wouldn't know the answer to a question, or I would go blank. Particularly as HRM was quite a new area for me when I began the PhD. I found the process of answering the questions really empowering and affirming, as it really made me understand that advice that 'you know your thesis better than anyone'. It felt like a really nice end to all of my work.

I got major corrections. It took them half an hour to talk this over and then they invited us all back in. My corrections were to be completed within six months. I believe this is a unique outcome for my particular university, where they have: no corrections/minor over three months/major over six months/revise and resubmit. I had a new job at this point and so it took me four months to complete this. The comments were that they were minor corrections, although the external examiner wanted me to rework one chapter (the introduction) in order to talk about HRM then HRD. I had originally done this, then changed it on the advice of my supervisor! Because it was reworking a chapter this then fell under 'major'. My minor corrections were adding in a sentence to explain a process more fully in my quantitative work and adding in one further reference of work done by my external examiner.

I felt fine in the viva and then anxious during the deliberation. Then, when given the result, a little sad to have major corrections, but also elated that I had got through the viva and passed.

One of my positive viva memories was the sense of pride I got from being able to answer all of my questions and demonstrating that I know this piece of work really well. Also telling my partner over the phone after I had passed. I was really crying and couldn't speak, but after a while I was able to calm down and say that I'd done it."

An experienced research fellow who completed her *PhD in Psychology &* *Mental Health* at a university in the UK between 2012 and 2018 shares her experiences:

"My PhD was in the field of Psychology & Mental Health and used a mixed-methods approach. This took around six years, as I completed this part time alongside researcher posts.

In order to prepare for my viva, I added chapter bookmarks so I could easily refer to a chapter if needed in the exam itself. I read over my thesis but found that I didn't need to do this too carefully as everything was still so fresh in my mind. I did, however, read up on some methodological issues which I anticipated could come up. I looked at my viva examiners' publications to help me think about the sorts of questions I might be asked. I chatted to my colleagues about what their viva had been like—some offered to send me questions which they'd been asked in their own exams, but much of my discussions were about what the experience itself was like (e.g. how their examiners had acted, how long it had lasted etc.)—I think I wanted to reassure myself that this wasn't going to be a horrendously scary experience (which it wasn't in the end)! Looking back, I think I did look up possible viva questions online but I'm not sure I found that too useful as it's rare you come across examples in your particular field.

I did not enjoy the build-up to my viva, I was very stressed leading up to it. I think it was because I had found the PhD quite stressful (particularly as it was part-time alongside work), so by the time I handed in I was exhausted. Also, I think I was also so close to the work that I wasn't sure how good or bad it was!

I chose not to have a mock viva. I did discuss this with my supervisors, but I think I was advised that for some students that had not been a useful experience—i.e. if it had not gone well, it had dented their confidence. I did, however, have a chat about the viva with my supervisors and they helped me to think about possible questions and answers. I was happy with this as I think I would have found a mock viva stressful.

Before my viva I felt extremely nervous, but I also had a feeling of relief—that I would finally get it over with. In my viva two examiners, an internal and external, and a chair were present. At the very start of my viva the examiners told me that they'd really enjoyed reading my thesis—this made me relax a huge amount and really helped my confidence. I believe my viva was relatively short—about 90 min I would estimate. I did not take any breaks—the time flew by!

During the viva I suppose at times I was surprised that the questions were quite straightforward; for example just asking me to explain how I'd done something (e.g. how did you go about synthesising the data). I had been considering all the 'difficult' questions that they may ask me, so I suppose it was a surprise (and a relief!) when these sorts of questions came up.

I suppose I did enjoy my viva! The examiners were lovely throughout and it felt good to discuss the work with people who were genuinely interested and who had expertise in my area.

I passed with minor corrections. We discussed the corrections in person when the examiners fed back their decision to me. They also sent a written list to me after the viva. It probably took around a day to sort out the corrections. My corrections included correcting some typos which I had picked up and some referencing errors. Fortunately, I didn't have to adjust any of the content of my thesis, which was a big relief! Afterwards, I felt more exhausted than I expected, but happy."

Emma Izon completed her PhD at the University of Manchester in 2020 and is currently a Trainee *Clinical Psychologist* at the University of Oxford. Here Emma shares her experience:

"My PhD at the University of Manchester focussed on the family environment for individuals with an at-risk mental state (ARMS) of developing psychosis. Individuals with an ARMS can experience distressing thoughts, feelings, and behaviours. This can include hearing or seeing things that others cannot, having unusual beliefs or distrustful thoughts. These experiences are typically reported during late adolescence or early adulthood, when individuals may still be living at home with their family or living with partners. These experiences may impact on the communication between family members and the home environment. My PhD involved a mixed-methods approach to understand the family environment and impact for family members/carers of individuals with an ARMS. My research involved synthesising and identifying gaps in the literature. I investigated changes in the family environment over time and the effect on symptoms (e.g., feelings of anxiety and depression) for both family members/carers and individuals with an ARMS. Finally, I explored family members'/carers' experiences and needs over time using semi-structured interviews.

I completed my PhD in 3.5 years full-time, whilst employed as a full-time research assistant on multiple randomised control clinical trials, at the Psychosis Research Unit, Greater Manchester NHS Foundation Trust. I needed good time management skills to balance both my PhD and my clinical NHS role. Monthly supervision was vital in allowing me to communicate any challenges I was experiencing and maintain motivation over the years. I was fortunate to have compassionate, supportive supervisors, who provided me great encouragement and belief along the way. I published seven papers from my PhD, which were all accepted in peer-reviewed academic journals.

To prepare for my viva, I instigated various strategies. As I was attending my viva online, I wanted to feel as comfortable and in control as possible. I made sure to prioritise my well-being in the weeks coming up—going for walks, being physically active and eating well. I spoke to my supervisors and other academics, who had been through the viva process, asking for advice and other tips. I did not have a mock viva. The month leading up to my viva, I read through my PhD and published journal articles, literature around the topic, and searched for any recent publications. I spoke to my partner, family, and friends about my research, which allowed me to practice discussing my research with diverse audiences. I also planned a small reward, no matter the result, to keep me motivated and congratulate myself for all my hard work.

I felt anxious and excited going into my viva. I had some inclinations about what the panel might ask, but, whatever the question, I felt ready to defend my work and justify my decisions.

The parties present during the viva included myself, and my two examiners, one internal (from my university) and one external (from another university in the UK). The examination lasted approximately two hours with no breaks. In the viva no questions surprised me; some were challenging, for example further expansion was required. Once I was able to relax, it felt less like a formal process, but rather a discussion between academics about my work.

Prior to my viva, my supervisors had mentioned this may be my only opportunity to discuss my research so intently with other people. I truly appreciated having a space to reflect on the process and share my contributions to the field with others.

As previously mentioned, I completed my viva online, as it was during the COVID-19 pandemic. After the panel had asked all their questions, I was asked to leave the virtual room. I was invited to return after 20 min, where I was awarded a pass with minor corrections. I was provided with a rationale in addition to proposed corrections and comments. Approximately a week later, these details were emailed to me. The corrections varied between expanding on points, and providing more rationale, to grammatical or spelling changes. I was provided a time scale to respond to these proposed changes.

My most positive viva memory was having the opportunity to discuss my research that I had created, tested, or explored and analysed with fellow academics and experience the feeling that I was an expert in the area.

Once I completed the viva and was awarded a pass, I felt a huge sense of relief. All my hard work and efforts had been worthwhile. I felt incredibly proud of myself and my accomplishment. It felt fantastic to share the news with my family, friends, and supervisors, who had supported me greatly over the years."

Carolina Campodonico completed her PhD in 2017 at the University of Manchester in the Division of *Psychology and Mental Health*, and now works as a lecturer in clinical psychology at a university in the north-west of the UK. Carolina shares her experience:

"In September 2017 I started my PhD in Manchester, and I did my viva in July 2021. To complete the PhD, I was awarded the Research Impact Scholarship, which was funded by the Division of Development and Alumni Relations from donations to the University of Manchester. The topic of my

PhD was positive adjustment after trauma in people experiencing psychosis, which means that I investigated the protective factors that shield against traumatic symptoms (PTSD and complex PTSD) and help people experience personal growth after going through a trauma (a concept know as post-traumatic growth).

To prepare for my viva I:

- I practiced—a lot—by doing mock vivas with many of my friends. I had a list of common questions and relative answers which I gave to my friends, and we role-played the viva multiple times.
- I read recent literature on the topic, to make sure I wasn't missing any important findings.
- I went through the publications of my examiners, to understand if there were topics on which they were particularly keen, as that might have influenced the type of questions they would ask.
- I re-read one chapter of my thesis every day, for the two weeks before the viva.

Before my viva I remember that I was stressed, my hands were sweaty and, overall, I thought I was going to pass out. I could only think "What if they ask me something to which I don't know the answer to?"

The viva was done online because of COVID-19, so it was me, the two examiners and an extra person who was there just to make sure there were no technical glitches (as per university regulations). It lasted exactly two hours.

I didn't have any breaks because I didn't want to, nor did I need them, but I was made aware at the beginning that I could ask for a break at any point. Doing the viva online was a great experience, and if I could choose again, I would choose to do the viva online rather than face to face. Having that type of "distance" helped me feel more at ease also because I knew I could quickly look for information in the thesis without having to scroll through the physical pages of the thesis.

I wouldn't say that I was surprised by any of the questions; however, I was surprised that we didn't cover the whole thesis. I had six chapters: an intro, a systematic review, a quantitative study, two qualitative studies and a conclusion. We spent some time on the intro, and some time on the systematic review, then we talked about the design/analysis of the quantitative chapter, followed by a couple of questions on the qualitative studies and nothing on the conclusion. I am not sure why this was the case; I like to think that they simply didn't have any comments on some of the chapters.

Surprisingly enough, my viva was a great experience. Both examiners were lovely and made every effort to make me feel comfortable. The external examiner noticed some trophies in my background and spent some time chatting about them at the beginning. Overall, it felt like having a nice chat. I didn't have the impression at all that I was being examined; it felt like a peer discussion among researchers. At some point I even (respectfully) disagreed with one of the comments of my external examiner, and he was totally fine about it; we just talked it through. There have also been a couple of occasions when we had a laugh. I passed with minor corrections, and, I mean, seriously minor (I could have done them in a couple of days if I wanted to). I had 25 changes, of which a good part were typos and errors in reference formatting. I also had two non-mandatory corrections, which I didn't know was an option. The corrections were sent by the internal examiner probably just a week after the viva. They were written in the same style in which the viva was conducted, i.e. the feedback was very supportive, and all the corrections has been discussed during the viva. It probably took me a month to address all comments, but simply because I was already working at the time, I was in no rush and, to be honest, I needed a break from the PhD. Some examples included typographical errors and clarifications:

1. P18. "While the conventional model of psychosis views delusions and hal-lucinations as indicators of underlying psychopathology (Read, Bentall, & Fosse, 2009)"—this is not the primary reference. Please use primary reference.
2. P.20—P20 four lines from the bottom—correct the citation, which isn't in APA style.

When the viva finished, I smiled. I literally just sat at my desk and smiled. It was like all the weight of the world had been lifted off my chest. I realise it sounds quite clichéd, but that's literally how I felt. Not excited, not like screaming, simply calm and content, like when floating in the sea!! My mind was so empty, but in a warm and comforting way.

My viva was the first time I felt like I owned my PhD: I was an expert in my field and the examiners where simply there to have a genuine conversation about my research."

Kamelia Harris completed her PhD at the University of Manchester in 2021 and now works as a project coordinator on a large clinical trial in the northwest of England assessing the effectiveness and acceptability of a suicide-focused talking therapy for people experiencing psychosis. Here Kamelia shares her viva experience:

"I started my PhD in 2016 at the University of Manchester, UK. The project was funded by the Mental Health Research UK (MHRUK) charity. *The topic of my PhD was understanding psychological resilience* to suicidal experiences in people with non-affective psychosis. I completed my PhD over four years and conducted two qualitative interview studies exploring people's experiences of resilience to suicidality, and a longitudinal study examining relationships between resilience, psychosis symptoms and suicide precursors (e.g., defeat, entrapment, hopelessness). Overall, 105 people with experiences of psychosis and suicidality took part in the studies.

To prepare for my viva I spoke to fellow PhD students who had had their viva about their experiences and tips for preparing for it. That helped me collate a list of possible viva questions early on in my training. I also practiced answering questions on my own and with others. I remember reading a book that a fellow PhD student lent me which was helpful. It's called *How to survive your viva* by Rowena Murray.

About six months before my viva, I attended Nathan Ryder's "Viva Survivors" workshop and got some useful preparation ideas. I would highly recommend it! I did not have a mock viva. Instead, I had a meeting with my supervisors closer to the viva date to discuss potential questions and how I could answer them. I got a useful tip from one of my supervisors which was to write a very brief outline of my thesis and bullet-point the aims, research questions, and key findings for each chapter. I felt excited and relatively calm before my viva and looked forward to talking to my examiners about my PhD in detail, albeit over Zoom.

My viva was conducted over Zoom due to social restrictions during the COVID-19 pandemic. I had an external and an internal examiner. The whole viva lasted two hours and we had a ten-minute break halfway through. I wouldn't say I was surprised by any of the questions that were asked in my viva. Overall, it was a very positive experience.

I passed my viva examination with minor corrections. I received these a few days after my viva in an email from the University's Doctoral Academy containing a document with a list of all corrections requested by the two examiners. It took me about four days to complete the corrections. My examiners then reviewed and approved the corrections before final submission. The corrections were primarily about describing the analysis method that I used in my qualitative study chapters in a bit more detail and providing the results of additional statistical analyses in my longitudinal study chapter.

As I had my viva in early 2021, during COVID lockdown, my post-viva experience was a bit underwhelming, to be fair, but I was happy that it went well. One of my key positive memories was when the examiners said 'Congratulations, you passed your viva!'"

A research associate who completed her *PhD in Clinical Psychology* at a UK university between 2015 and 2022 shares her experience of preparing for the viva and the viva:

"My PhD was in Psychology. The aim of my PhD research was to develop a behaviour change intervention. The development of the intervention was evidence-based and informed by psychological theory.

It took me six years to complete my PhD. My PhD was a four-year funded studentship, but I had an interruption period for maternity leave and I returned to the PhD part-time.

I prepared for my PhD viva in a few ways. I presented my PhD research at a seminar in my department. The seminar was an opportunity for me to speak about the PhD study and to answer questions about it. I had asked a colleague to note down all the questions I was asked for me to be able to reflect on how I could improve my answers. I spoke to colleagues who had completed their PhDs recently. They gave me examples of the types of questions they were asked. I put all of the information together and prepared answers for these questions. The university I attended also provided some generic questions for viva preparation and I used this to test myself on how I would answer these questions. I arranged a mock viva with a senior staff member in my department.

I was incredibly nervous going into the viva. In my viva, there was an internal and an external examiner and it lasted three hours. I did not take a break but was informed at the start of the viva that I could request a break at any time. I was not surprised by the questions. I did not enjoy my viva at all. I found the experience very difficult. I was relieved it was over.

I received major corrections which I received in writing. This was very helpful to ensure I addressed all of the points the examiners raised. Initially, I was given six months to complete the corrections. Due to the impact of the COVID-19 pandemic, I needed an extension to complete the corrections. I was given a six month extension. The majority of my corrections were about adding more detail to the thesis. I had written the thesis in alternative format, but in doing so, I lost sight of the level of detail needed in the introductory chapters. For example, I was asked to include more information on the justification of my research methods at the beginning of the thesis.

The corrections I had to make have undoubtedly improved my thesis. I also feel that this experience has helped me to become a more diligent and empathetic academic, in particular for when I am a supervisor or examiner to university students."

Liz McManus completed her *PhD in Psychology* at the University of Manchester 2016–2021. Liz currently works as a lecturer for the BSc Psychology Programme at the University of Manchester. Liz shares her viva experience:

"My PhD started in September 2016 and was due to be a four-year course funded as part of the Biotechnology and Biological Sciences Research Council at The University of Manchester in the division of Neuroscience and Experimental Psychology. Due to the COVID-19 pandemic during my final year, I had a six-month extension, meaning I submitted my final thesis in March 2021 and completed my viva in May 2021. My research focuses on the effects of social forms of stress on the brain and how this may impact cognitive abilities such as memory. My programme was part of a doctoral training partnership, the first six months of my PhD were 'lab rotations', during which I completed three, two month mini-projects with each of my supervisors as an opportunity to prepare me for my later work. My programme also supported students to complete a three-month professional internship placement, away from our PhD work. The purpose of these placements was to allow us to explore fields outside of our area of study. For my placement I worked in collaboration with Taipei Medical University to develop an analysis tool to link neuroimaging measures to regional receptor densities.

During my viva preparation, here in the UK we were starting to come out of the third lockdown, so everyone was still working from home. I wish I'd had the opportunity to be in the office and chat general viva things with other PhD students while preparing, just as a form of reassurance, motivation, and shared nerves. After handing my thesis in, I very deliberately avoided re-reading it for about a month so I could come back to it with fresh eyes as I'd spent so long reading it while preparing it to be submitted. About a month before my viva I started re-reading my thesis and making flashcards to summaries each chapter, the overarching 'story' or 'golden thread' and for key mechanisms/theories that I thought would be important. These flashcards helped me condense my knowledge into concise points that I could then build on further during the viva if I needed to. They were more prompts than exact details. Planning out and creating the flashcard for my overarching thesis story was very useful though as it helped me solidify my narrative of how my chapters linked together to create a whole thesis rather than just distinct papers or experiments.

My supervisors did a mock viva with me a couple of weeks before my real viva. Although this was a great opportunity for them to quiz me on my research and the motivations behind certain decisions and interpretations, it did highlight one very important things: A viva examiner (whether real or mock) will ask you questions based on their background and interests. For me, my supervisors were much more focused on neuroimaging and the brain, so many of the questions in my mock viva focused on these elements of my thesis. But my examiners were more interested in the cognition and biological elements and asked questions far more centred around that and were much less interested in the neuroimaging side of things. So top tip: know what your examiners are interested in; this is the perspective they'll ask questions from and get your mock examiners to think like them!

I was surprisingly less nervous than I expected to be on the morning of my viva. I think it's something all PhD students are anxious about for a very long time and when it finally arrives it's more of a relief that it's finally here.

My viva was via Zoom (COVID) and was with myself, my internal examiner and my external examiner. Many people often have a chairperson during their viva as an independent observer, but given how experienced my internal examiner was and that I was comfortable that they would keep things on track and unbiased, I didn't feel the need for a chairperson in my viva.

My viva in total was around three and a half hours long. At around 5 min into my viva, the internet at my house failed and I lost the Zoom connection. After a frantic 5 min running around the house trying to fix the internet, I had to rejoin the Zoom call from my phone using mobile data. The remainder of the viva (well over three hours) was done on my phone. Luckily, the battery and signal didn't fail and the calls connection was good throughout. Once I was back online the examiners were very kind about the situation and continued with the viva as normal. They could see me, so to them it wasn't particularly disrupted. But for me, not only was it unusual anyway to be doing my viva remotely, but to now be doing it on my very small phone screen, not my computer (with two screens!) was incredibly stressful and rather surreal. Because the internet wasn't working, I couldn't access the digital copy of my thesis as it was stored on the university server so I needed internet to access it. Luckily I had a printed copy of it with me that I'd made notes in, so I could still flick through to certain pages where necessary to clarify things for my examiners. After about the two-hour mark, random notifications and texts kept flashing up on my phone screen (while I'm trying to answer the examiners' questions!) from my supervisors, friends and family asking how the viva had gone and if I was finished yet. Once we'd gone through my whole thesis, the examiners put me in the Zoom waiting room so they could discuss and deliberate and then let me back into the virtual room after around 10 or 15 min.

We took breaks approximately every hour or so during my viva. The time went really quickly but the breaks definitely helped things feel less intense and gave me the opportunity to refresh and relax briefly.

Knowing the backgrounds of my examiners, I wasn't surprised by the approach the examiners took to questions. I was a little more surprised at the importance they places on some points that I myself would have considered more minor, yet they were less interested in certain aspects that, to me, were much more important/interesting. Again though, this reflects the difference in our areas of interest rather than anything unfair or too unexpected. Overall, the examiners were not trying to 'catch me out' with their questions. Some were probing alternative ways of doing things to gauge my understanding of alternative methodologies and the benefits of the methods I had chosen compared to these alternatives.

My examiners joked with me at the time that the circumstances of my viva would one day make for a funny story, especially given that my thesis was literally about the impact of stressful situations on cognition and memory. My research is very relatable and understandable from a lay perspective. So, when giving any talks about my work, I do often joke about how I might forget what I'm about to say, because I'm stressed, and that was my research topic at work in real time. But at the time, after four and a half years of work and a global pandemic, I wasn't ready to see the funny side at the time. The situation was very stressful for me, and I was worried that I was going to fail and that all my hard work was about to be derailed by a bad internet connection. But as the viva went on, I settled into the unusual set-up. Although it was challenging, my examiners were very fair and it was actually a good opportunity to have an academic discussion with other people about my work and its pros and cons and the possibilities for future work.

I passed my viva with minor corrections. These were mainly clarifications on some points, a minor terminology change and some typos. The corrections came through about two weeks after my viva and I then had a month to complete them. Having a short break between the viva and receiving the corrections was useful as it gave me time to reflect on the viva and mentally prepare myself to work on the corrections rather than feeling mentally fatigued with it. One example of a correction I was given was to clarify my age ranges in one of my papers. This paper discussed stress experienced during childhood and adulthood and the examiners asked for further clarification on the age range these times best covered.

Once it was done, I was exhausted and very relieved it was over. It took quite a while to sink in that my PhD was finished (pending corrections). It was frustrating that because of the pandemic, the celebrations that I had enjoyed for friends' vivas (champagne in the office, dinner and drinks etc.) weren't possible for me as we were still in lockdown/restrictions. This really emphasised the importance of celebrating the things you work hard for during your PhD with people you have worked with and want to share those moment with.

Although I wouldn't have agreed at the time, the fact my internet broke and I did my viva from my phone is finally the funny story I was promised it would be. It epitomises the split-second changes, adaptations and generally weird circumstances that happen in everyone's PhD, especially those of us who were PhD students during COVID."

Brioney Gee completed her *PhD in Clinical Psychology between* 2013 and 2016 at a university in the east of the UK. Brioney shares her experiences:

"My PhD was focused on a set of symptoms experienced by some individuals with psychosis (called negative symptoms). The aim was to try and better understand the psychological factors that are involved in causing and maintaining these symptoms. I completed it between October 2013 and December 2016 at a university in the East of England.

My PhD took three years. My research was based on secondary analysis of a pre-existing data set from a large-scale study I had been involved in prior to beginning my PhD. Not having to collect primary data made a big difference to how quickly I was able to complete.

I prepared for my viva through re-reading my thesis, taking notes on questions that I thought might be asked and marking up the pages for easy reference during the viva. I then wrote out answers for the questions I thought might be asked and practiced saying them aloud to myself initially, and then to friends and family. I also had a mock viva with one of my supervisors a few weeks before the real thing. I don't think it bore much resemblance to the real viva but helped ease my nerves to some extent.

Before my viva I felt like I was preparing to have a kidney removed without anaesthetic.

In my viva I had two external examiners and an internal chair. It lasted around 1.5 h, with no breaks. There were one or two questions I hadn't prepared for, but nothing completely left field. I was pleasantly surprised that the things I had worried they would raise didn't come up!

I am, however, not sure I'd say I enjoyed it; enjoy is not quite the right word, but it was a positive experience overall. I was very nervous initially but felt much calmer after it became clear that the examiners weren't about to rip me to shreds! I'd heard lots of people say this, but it was true that it felt good to discuss something you'd been working on for so long with people who had read and understood it, and who seemed genuinely interested in what you had learnt.

I passed with no corrections (although I did fix some typos before handing in the final version). I think the fact that one of my examiners was Dutch played a big role in my not being given any corrections as (from what I understand) in the Netherlands the viva is more ceremonial and it's unusual to ask for corrections at that stage. After the viva I mostly felt relieved but also pretty proud of myself.

The most positive moment was definitely when I was called in after the examiners had finished deliberating and was told that I'd passed without corrections."

Anamaria Churchman, a Lecturer in Psychology at a university in the north-west of the UK completed her PhD in 2019 at the University of Manchester. Here Ana shares her experience:

"*My PhD explored school-based psychological interventions* for young people (11–16 years old) experiencing psychological distress. The title of my PhD thesis was 'The feasibility and acceptability of a PCT-informed psychological intervention for young people in a school setting'.

Considering half of all mental health problems start in adolescence and access to appropriate and effective support for this age group is very poor; the PhD was a great challenge. The project initially focused on offering young people a novel psychological intervention called Method of Levels (MOL). MOL had not previously been used with young people, as such, the first part of the project sought to understand how young people experience it and if it had the potential to be useful for young people.

During adolescence, a large part of young people's distress arises as a result of friction in the parent–child relationship which affects young people's well-being. Consequently, the second part of the project focused on developing a parent–child activity that could be used alongside individual support for young people.

The final part of the project proposed the use of the two-component intervention to support young people. The parent–child activity invited parents to work with young people to address parent–child conflict and reduce emerging mental health difficulties. Alongside this, young people were offered one to one support in the form of MOL therapy. The full programme of research generated six publishable chapters, taking the final thesis to nine chapters including the introduction, methods and discussion.

My PhD involved both clinical work (training and delivering MOL therapy) and scientific research (collecting and analysing data). Most of my time was spent in a secondary school in Manchester, UK where I worked closely with young people and their parents.

My PhD took three years to complete. It began in September 2016 and I submitted my thesis in September 2019. The viva took place in November 2019 and I graduated in December 2019.

I first learnt about the 'viva' process when I attended on introduction course ('Life as a PhD student') in my first year of post doctorate study. The course was organised by the university and was designed to give us an overview of what to expect and what to avoid during our doctoral studies. The final slides referred to 'viva' and gave us examples of typical questions asked. I'm originally from Romania, so the only viva I knew was the yummy chocolate pillow snacks (same spelling, different pronunciation). I was not familiar with the examination system and walked out of that course extremely confused. A quick Google search cleared things up shortly after that.

As my studies were coming to an end, I sought to prepare myself as much as I could for something that I have never quite experienced before. The university ran a number of workshops aimed at equipping us to successfully write up the thesis and pass the viva. However, these were badly timed, with sessions running in April/May when I was in the thick of the final write-up and not really in the right mindset to worry about the viva. By the time my viva was due in November 2019, my scribbled notes from earlier workshops did not make any sense so I sought further resources online. I stumbled across a 'viva toolkit' made available by the university. The folder contained a few general documents that helped me understand the process (how the assessors mark you prior and during your viva) as well as the PowerPoint slides ('Surviving your Viva') for the workshop I attended earlier in the year. Prior to reading these documents, I felt quite confident and thought that I had already done the hardest part—I had written and submitted the thesis—the viva was just a formality. However, after going through the 'viva toolkit' documents, I suddenly became aware of the scary unknown that was awaiting me.

The realisation of how little I knew about the viva process began to sink in and bring panic and dread. I ran back to the PowerPoint slide, which ironically was entitled 'Surviving your Viva'. But unfortunately, refreshing myself of the contents of the slides, I did not feel equipped to survive my viva. I was left with a vast number of questions and I was thanking the Lord, I still had a few weeks before my viva. I was desperately searching and looking for more resources that could equip me, but to my utmost disbelief the literature was scarce. At the end of the 'encouraging' PowerPoint slides ('surviving your viva'), there were a couple of book recommendations and a website link—surely there was more written on the topic—but no, resources were scarce. What I found really helpful was talking to a couple of colleagues who had completed their PhD and had gone through their viva exam. Their accounts were real and raw. This experience was going to be challenging but rewarding. Their best advice was read through your thesis and make sure you own each statement.

I did not have a mock viva. I thought my supervisors would struggle to be impartial or tough given that the project was a collaboration and they had been part of the journey. I did ask my supervisors to be critical and point out any flaws and weaknesses they recognised in the thesis. This helped me think critically about ways I might use to defend the areas of the thesis that did not come across as strong or clear. I also found it helpful to ask colleagues who have passed their viva about any challenging questions that they struggled with or found difficult to answer.

I was definitely nervous going in to the viva. I knew it was going to be tough, but I was confident in the work I had done and the quality of it. The imposter syndrome followed me from the time I began until the time I submitted the thesis. Was I doing something that was good enough? Will the amount and quality be enough to pass? But, as I walked into that room to defend my thesis, I knew I could stand by and argue every decision that was made and explain how things were done.

During my viva, I had two examiners present in the room. I did not know my internal examiner as he was from a different department (education). This made the process more daunting. However, I enjoyed reading about his background and research prior to my viva. I knew that he will have read my thesis through the lenses of his experience and research and as such tried to think what questions he could pose.

I walked into the viva examination knowing it was going to be challenging, but some questions did surprise me and were extremely difficult to answer. But the thing I was most surprised about was the sinking feeling I got when I was asked some tough questions. As I tried my best to answer them, one of the assessors was shaking their head and tutting. I did my best to stay focused and remember that because all questions were about the work I carried out, I had the skills and ability to answer them. I was the expert in my area. So, that meant that I could share with them what I knew and what I found through my research.

My viva lasted just under two hours. It started at 14:00 and around 15:50 I was asked to leave the room and come back in a few minutes. I must admit it did not feel that long at all. Because it was intense and challenging, time flew. I did not take any breaks during my viva. I was informed that I could take a break but time went so quickly, I did not feel I needed to. I was keen to get through all the questions they had and have the ordeal over with, rather than prolong it any longer.

It would not be fair to say I enjoyed it, as it was a difficult and challenging experience. I can definitely say it was rewarding and cathartic. After three years of laborious work, with many ups and downs, with different challenges (mainly around managing the supervisory relationship), I could finally wrap it all up with a beautiful bow (bound thesis) and present it as a labour of love to two strangers who were curious and interested in the work I carried out.

The outcome of my viva was minor corrections, with five points to address and a few typos.

I received my corrections two days after my viva in an electronic letter. The guidelines suggested that I was required to submit the corrections within four weeks of receiving the letter. However, I was keen to make it to the winter graduation which was scheduled for 9 December. In order to qualify for this, I was informed that I needed to submit and have the corrections approved by the internal examiner before the 27th November. My viva took place on 13 November. Due to the minor points raised, I successfully completed and submitted my corrections within a week of receiving the results letter. These were quickly approved and I was able to attend the winter graduation shortly after. What a whirlwind experience.

I received five minor corrections. These related to clarifying a few things and adding an overall conclusion in the last chapter. A few examples of the specific corrections include:

- A reworking of the discussion of the scheme on page 49—add some commentary that explains what the PhD was as a whole, an assessment of feasibility and acceptability. Please ensure that the specific grounds for the acceptability are defined within the context of this research.
- Further situate the author in the context of the school—(suggest to include in Chap. 2) e.g. a reflection of the position of the researcher and the school. Environment and context and how this might influence the research and its findings.
- In the concluding chapter, include a statement as to the overall/overarching contribution of the thesis. This statement should encapsulate the thesis as a whole, rather than being viewed by chapter.

After the viva was finished, it felt like a blur. I walked back to my office, where a few colleagues were keen to find out how it went and the result. They had prepared a card and a cake which was extremely kind (and optimistic) of them. I felt overwhelmed. It was over but it did not feel that way. The physical tension in my body was just about easing off. I could let my shoulders down. Is this the end of three years' worth of work? What does that mean? What next? For the rest of the evening it felt like I was floating, not walking. It did not feel real. I had to keep reminding myself—it is over. But it was not quite over—there were still corrections to do. The bitter-sweet rollercoaster of 'completing' a PhD.

The elated moment when I was told I passed with minor corrections. You underestimate the power those few words have on you. You think you have done a good job, you think you have defended it well, you are confident, you try and stay strong, composed throughout the whole time you are in there. But the words spoken over you when you are invited back in the room (to have the final result communicated) have the power to make you or break you. My world changed when I heard those precious words— You are now a Doctor in your field. You have earned the highest academic qualification. I know that message brings with it responsibility and I hope to live up to it."

Emily Eisner completed her *PhD in Clinical Psychology* at the University of Manchester from 2011 to 2019. Here Emily shares her experience:

"My PhD was in clinical psychology. More specifically, I looked at how we can help people who experience psychosis to predict whether they might soon become unwell again (relapse). I examined whether adding 'basic symptoms' to standard assessments of early signs of relapse might improve prediction. I also examine the feasibility, acceptability and validity of using a smartphone app (ExPRESS) to assess these hypothesised relapse predictors. I completed the PhD at the University of Manchester (UK), passing my viva with no corrections in November 2019. As I took so long to do my PhD, I had published all five papers before submitting the thesis. My institution allows us to submit in 'alternative format' which worked well in my case.

My PhD took eight years from start date (2011) to end date (2019). This was the equivalent of four years full-time. For the first three years I was full-time, and then I was part-time from 2014 to 2016 and from 2017 to 2019. I had two periods of maternity leave during my PhD, 2013–2014 and 2016–2017. My PhD was funded by the Medical Research Council.

To prepare for my viva I did the following:

- Read through my thesis, making notes in the margins of things I would ask if I was reading it for the first time.
- Made sure I had a really firm grasp of the overall 'story' of the thesis and how it fits with the bigger picture.
- Prepared for obvious introductory questions, methods questions, etc. using lists of questions that other people had been asked in their vivas. My institution provided some lists of previous people's questions/experiences.

- Read through the viva report of someone else who had previously had the same examiners and looked for questions that could come up in mine where our topics overlapped enough to be relevant.
- Read anything relevant that either of my examiners (especially the external) had published recently. I actually started doing this while I was still writing the thesis (as soon as I knew who my examiners were likely to be) to ensure that I was citing anything relevant from the examiners in the thesis itself. This came pretty naturally as their research was very relevant to the topic of my thesis.
- In retrospect, this was over the top, but I prepared an answer to why I had used 'data is' rather than 'data are' throughout my thesis! I was not asked about this in my viva!
- The morning of the viva I read through my bullet point notes one last time, went for a run, had a shower and dressed in some comfortable but relatively smart clothes and felt ready to go.

I did not have a mock viva. I actually felt very ready for the viva. I was slightly nervous but also looking forward to getting the viva done.

In my viva, there was an internal examiner and an external examiner present. I cannot recall how long it lasted exactly but maybe 1.5 h to 2 h. I did not take any breaks during my viva. As far as I can remember, I was not surprised by any questions. I felt that I had prepared well for the viva, and I didn't feel at any point like the examiners were trying to 'catch me out'. All their questions seemed reasonable and along the lines that I expected. I oddly enjoyed my viva! Partly I think this was because I had invested so much time (eight years) to complete the PhD and it was actually really nice to spend time discussing it with two clinical academics whose opinion mattered to me.

I passed with no corrections. Afterwards it was a big relief to have finally finished the whole PhD process. I enjoyed a celebratory drink with my examiners and supervisors and then, of course, celebrating with friends and family. I was excited to finally call myself Dr. after eight years of hard work!"

Neuroscience

Joanne Sharpe completed her *PhD in Neuroscience* at university in Manchester, UK in 2017. Her PhD was funded by the Medical Research Council. Currently, Joanne is a post-doctoral researcher at the University of Sheffield. Here Joanne shares her experiences.

"I started a PhD in Neuroscience in Manchester, UK in September 2017 and finished in September 2021. I used fruit flies as a model organism to investigate mechanisms of motor neurone disease and frontotemporal dementia. It took 3.5 years, plus a six-month extension to account for COVID-19 disruption.

To prepare for my viva exam I looked through my thesis and ensured that I understood the rationale for each experiment, and there were no terms that I couldn't define if asked. I also read around the key themes and made mind maps to help me remember important points. I did not have a mock viva. However, I was asked to present a 15-minute talk in my viva so I practiced this in front of my lab group.

Before my viva I was very nervous, but I knew that I was prepared and no one knew my project as well as I do. My internal examiner and my external examiner were present and it was held via Zoom due to COVID-19. It was precisely three hours long. I was encouraged to take breaks, but I didn't end up needing any. At the two-hour mark my examiners asked if I wanted a break but I chose to keep going.

I think the hardest questions for me were on details I had referenced in my introduction that I struggled to explain fully because it was background that was outside my field of expertise (genome-wide association studies). The rest of the viva contained no surprising questions; they were all reasonable!

Personally, I don't have the confidence to enjoy my viva fully, because I felt the pressure and nerves of the occasion. However, it was an enjoyable discussion, and my examiners were fair. It was also probably the only time I will get to talk about my own research for so long, so you have to savour it!

I passed with minor corrections. I was asked to split up some figures, add a couple of more recent references and correct minor typos. I was given a marked-up copy of my thesis. It took me a couple of days to do it.

Afterwards I felt happy, but it was quite anticlimactic. It was strange that I didn't get a great deal of feedback; once it was over, they said I did well and would get minor corrections, and it was over.

One of my positive viva memories was having an intense discussion about a complicated biological process and I was able to completely justify and explain myself and the examiners ended up agreeing, which was very satisfying."

Emily Hird completed her *PhD in Cognitive Neuroscience* at the University of Manchester between 2013 and 2017, funded by the Medical Research Council. Emily shares her experience:

"My PhD investigated the brain and cognitive mechanisms underlying the placebo effect in pain. I began in October 2013 and submitted in November 2017. My viva was in December 2017, and I submitted my final thesis (with minor changes made) in February 2018.

To prepare for my viva I took some time off after submitting in November (very important as my mind was saturated and I was very tired by then!). After a week or two of relaxing, I re-read the thesis carefully, and formulated a list of questions in different themes (statistics, research design, literature etc.). My supervisor also kindly sent me a set of questions that might be asked. I answered as many questions as I could think of, and I practised them many times. I also bought a new outfit for the viva which helped me to feel more confident and professional.

I had a mock viva with my supervisor, which was helpful. Also, we did have a mock viva at the end of the first year of the PhD, so I had some idea of what the process would be like.

Before my viva I was nervous, but ready to get the viva done. I was not sure what to expect but felt reasonably prepared. In my viva the following individuals were present, the main examiner, who came over from a nearby city, the internal examiner, who was from the department, and a chair, who oversaw the viva because the internal examiner had never examined a viva before. I think it was about three hours. I do not remember taking any breaks.

The first question took me completely by surprise—it was "What is your favourite bit of the thesis?" This was meant as a bit of an icebreaker, but I had been expecting to be challenged, so this was difficult to answer in the moment! I was also surprised by the fact that the chapter I was most worried about was the chapter with the fewest questions. In the end, this chapter was the most interesting output from my PhD, and a good experiment, but at the time, I felt really unsure about it. Some of the questions focused on my writing, which I hadn't expected. But these were easy to answer as they weren't about the fundamentals of my research. These were minor corrections. At one point the examiner asked me if I'd heard of a theory quite relevant to my thesis, and I hadn't. I just said I hadn't, and we moved on. In general, the questions were mixed—some I had anticipated, and some I hadn't. I just stayed cool and answered carefully without rushing.

Upon reflection I did actually enjoy my viva. The build-up was not enjoyable but the viva itself was a great experience. It was the culmination of four years of hard work and allowed me to share and discuss my research with other experts.

I passed with minor corrections. I received my corrections via a set of notes, that my internal examiner kindly sent to me after the viva. I completed them within two months, whilst I was on a research placement. Lots of the corrections were around writing style. Can't remember otherwise. After my viva I was tired and very happy that it was done. I was proud of myself! The feeling of being told that I had passed was incomparable! Also, my supervisor arranged and hosted a celebration that evening, which was really nice.

One of my positive viva memories was being told that I had passed. And then telling my supervisor! Knowing that I had passed from student to Dr., after four years of working towards this, was exhilarating."

A researcher who completed his PhD in the USA shares his experience.

"*I did my PhD in Cognitive Neuroscience* in the latter half of the 1990s at a well-known university in New England. My research was on how the human nervous system manages combinations of rotations in multi-joint aimed movements. I took five years, which is the standard timeframe in that system.

My viva had a public format. I had to first give a talk on the main purpose and findings of my thesis to an audience consisting of my committee (by supervisor and two other academics who also served as the examiners), other staff and PhD students from the department, and invited academics from other departments in my university and from other universities in the region. After the talk, there was a period of questioning by members of the committee, then invited guests, then the rest of the audience and then back to the committee, and so on. Following this, the committee and all attendees with a PhD had a closed meeting while the rest of us waited outside. Having made their decision, the committee came out to announce the outcome.

My preparation started with making sure that my talk was an accurate reflection of the main things learned from the project. I also made a list of limitations that could be the focus of questions and prepared how I would address them. The same for the implications because the audience containing people without specific domain knowledge tend to ask more questions about implications than project details. As I expected, I was able to pre-empt all the questions about the specifics of the project itself. Some of the wider audience questions were not entirely predictable, so some thinking needed to be done on the spot.

I did not have a mock viva, but there were opportunities to give talks on the project at various points along the way, so I had some practice of presenting the topic and answering questions about it.

My talk was about 45 min, I think, and the Q&A was about 90 min. I did not take any breaks. The whole thing was one session. I could have taken a comfort break between my talk and the Q&A but I declined.

I was not surprised by any of the questions addressing the project's work specifically. Some of the questions from the general audience were a bit surprising. One senior professor simply asked 'What about golf?' I had to think on the spot what the implications of my findings about the degrees of freedom of arm movements were for skilled golfing.

I did enjoy my viva. I had very strong knowledge of the topic area and so I had few concerns about not being able to answer the questions.

I received minor corrections. These were requests to explain some things in more detail. I was given the list of corrections within two days of the viva. It took only a few days to do the revision work, but I had to spread it over a couple of weeks as I had to travel during that period. They asked me to better explain the maths I used to represent rotations. Much of the previous literature I drew on studied eye movements. I was asked to better explain how eye and arm movements related in terms of managing multi-axis rotations. I was asked to expand a theoretical discussion at the heart of the project—the claim that the nervous system attempts to reduce the degrees of freedom of controlled movements.

Once my initial presentation was ready, I was quite calm and ready before the viva started. Even though I didn't have a tough time during the viva, I do remember a great sense of relief when it was over. I was already in my academic post when I defended my thesis, so I didn't quite feel the occasion as marking the end of one life stage and the start of another. This is very much a part of the experience for most people.

My most positive viva memory is the opportunity publicly present and discuss a fundamental and interesting characteristic of the nervous system that is still not well understood."

4.3 Anthropology

Charlie Rumsby completed her PhD in 2019 at the Centre for Trust, Peace, Social Relations at Coventry University in the UK. She is currently the Sociological Review Fellow at Keele University and a Visiting Research Fellow in Anthropology at the London School of Economics and Political Science.

"I completed my PhD in Development Studies (2014–2019) at the Centre for Trust, Peace and Social Relations, Coventry University, United Kingdom. *My PhD was an interdisciplinary study, with an anthropological undercurrent.* My thesis explored modes of identity and belonging among noncitizen children of ethnic Vietnamese decent in Cambodia.

My PhD was completed over five years. I did nine months ethnographic fieldwork in Cambodia, and I took nine months maternity leave during that period.

To prepare for my viva exam I Googled "Top questions that are asked in a viva", and then I set about annotating answers to those questions. I also had several telephone calls with my director of studies who coached me and asked me questions, albeit in an informal way. I was fortunate that I could set time aside in the working day to prepare for the viva. Although, it was no more than around 20 h in total. I also had a mock viva. My mock viva was delivered by colleagues also completing their PhDs.

Before the viva I was nervous. My director of studies took me for coffee and kindly told me stories unrelated to my research to keep my mind busy. In my viva, an external and internal examiner were present. My director of studies and a chair were also present. It lasted two and a half hours with no breaks.

I did appreciate the opening question: "What did you learn about children during your research?" It offered a chance to humanise and ground the viva. We were discussing real people whose lives matter. It was a privilege to share the stories of the research participants in the viva. There were moments when key ideas in my thesis were challenged. I was not sure at the time whether this meant the examiners did not like them, or not. However, where I felt their comments/ questions on my theoretical choices added to my thinking, I would reply that I would take their suggestions through to publication when the time arose.

I have spoken to people who find the moments of challenge/questioning in the viva an intimidating experience. It is hard going. Yet, I encourage people who are due a viva to think about how their examiners' thoughts/questions can be applied later in their academic publishing. The viva, as I understood it, is not about assessing a finished article. It is about assessing a piece of time-bound research. Perhaps students would feel less intimidated if they were coached this way. I was coached by my director of studies. It helped me so much.

I really enjoyed the viva. I tried to avoid seeing the viva as a grilling and an event to be terrified of. I was nervous for sure. Yet I knew going into my viva that I was the expert of experience; I was the one who had carried out the research that subject experts had read. I was confident I could defend the choices I had made in the research design and analysis. I was also very open to receiving feedback and processing the research with people whose work I highly regarded. I knew the viva would be a once-in-a-lifetime opportunity. So, I tried to get the most out of it.

I passed with no corrections.

The most positive memory was the end. I waited outside for the internal decision to be made, and I thought I had done enough to have minor instead of major corrections. To be told I had no corrections was a joy. I was genuinely surprised. I felt my hard work, and the risks I had taken to make theoretical suggestions in my thesis, was affirmed."

Mathematics

Nicole Pearcy completed her *PhD in mathematics* between 2011 and 2015.

"I started my PhD in 2011, it took four years to complete. The aim of my project was to use techniques from network science to develop novel computational approaches for relating the topological organisation of a variety of bacteria to the environmental pressures they evolved in.

I first prepared for common viva questions, such as 'summarise your main findings' so that I had confident answers ready in advance. I then went through my thesis and prepared answers to questions that I would find challenging to answer. I did a mock viva.

My external and internal examiners and an independent chair were present in my viva. I had a very short viva—it was only around 2.5 h long. I did not take any breaks, but the examiners said that we could have a break if I needed one.

I don't recall being surprised by any of the questions. There were a couple of questions I couldn't answer, but these were questions about related work and the examiners reassured me it was fine.

I enjoyed the viva much more than I expected. It felt like I was having an interesting discussion with colleagues, rather than being grilled by examiners.

My outcome was minor corrections. The examiner wrote his questions/corrections in the back of my thesis. I took around six weeks to finish the corrections because I was working as a post-doc in a new group. The actual corrections only took around a week to do."

Multidisciplinary

Theodore Hughes-Riley completed his *multidisciplinary PhD* between 2010 and 2013.

"My research was highly multidisciplinary and focussed on the development of a novel contrast agent for taking magnetic resonance images of the lungs: The main subject areas covered were physical chemistry, physics, engineering, and medicine.

Between starting my studies and graduating it took me just under five years, however the bulk of my PhD, including all of the experimental work, took three years and three months. At this stage I also had the majority of my thesis drafted. I believe that I submitted after just over three-and-a-half years.

To prepare for the viva, I spent just over a week carefully revising the content of my thesis. There was a fairly large gap between my submission and the viva itself, so I had forgotten a lot of the details. I was in full-time employment when my viva took place, so I took a week off as leave to both revise the thesis, and to try and mentally prepare myself.

I did not have a mock viva. It was tricky finding the time because of my job, so I did not ask to have one: With hindsight I think that it would have been wise to have a mock viva.

I think my viva was around two hours long. I do not recall taking a break during the viva.

The only people present where me, my internal examiner, and my external examiner.

The thing that surprised me was the strong focus on one of my three experimental chapters. Almost all of the questions were either on my theory chapter, or this. My expectation at the time had been that the questioning would be more spread across all of the thesis, which wasn't the case. It is worth noting that the chapter that was the focus of a lot of the questioning wasn't my longest experimental chapter.

I remember before and after the viva fairly well, but I don't remember a lot of the viva itself. I was definitely nervous. I think that because you are constantly being asked questions I really just focussed on what I was having to address at the time.

I received minor corrections, with three months to complete them. I seem to think that I received my corrections via e-mail. I recall getting these fairly quickly. It took me around seven full days to complete the corrections, however I did not do these continuously and worked on the corrections over the space of about a month. After the viva there was talk about me having to do a couple of extra experiments however it transpired that the extra data requested had already been collected by a colleague (I made it clear in the thesis that they had collected this). Otherwise, I think that the most major changes were adding some extra paragraphs to explain some things more clearly.

Medicine

Barney Scholefield is a *paediatric intensive care consultant* at a large hospital in the U.K. Barney completed his PhD in the UK in 2012. Barney shares his experiences:

"My PhD was very practical. The title was 'The feasibility of performing a randomised controlled trial of therapeutic hypothermia for neuroprotection after paediatric cardiac arrest in the UK'. It took 90,000 words to come up with an answer! It was very much data science, clinical database, some observational studies. There was nothing to do with the lab.

It included some systematic review work, quite a lot of statistics and data crunching but in essence I was looking at the use of therapeutic hypothermia in children on the paediatric intensive care unit.

I was based at a university, but I was very much based at the Children's Hospital. This was because in addition to my PhD I was also working clinically on the paediatric intensive care unit. It is often challenging for medical clinicians to do a PhD, especially as a registrar as you progress through your training, as you often have a family and a mortgage. My wife and I had our children whilst I was doing my PhD.

The structure of my PhD was a big introduction, and then each chapter itself was a project/paper. There was no separate methods chapter. When I submitted my PhD, I think there were three papers that had been published. Two of them were from one chapter, one was from another, so there were two surveys that I published and one systematic review. And I think after the PhD I then published two of the other major data chapters. Five papers came out in total.

To prepare for my viva I did the normal work, reading through my PhD from cover to cover and putting sticky notes on the pages and trying to think what it is that people may ask me. I also had ideas about the weaknesses and holes that the examiners may question me about. I also tried talking to my supervisors to get some sort of insight into what the process would be. My supervisor was particularly unhelpful, because one, he was very clever, and two, he did an MD rather than a PhD. My supervisor also had published everything that went in his MD, when he submitted his thesis, so he got a letter back saying, there's absolutely no reason to viva you, you can just have the MD. So, he didn't actually have a viva. I was faced with this idea that I was already lesser than my main supervisor.

I think like any interview and exam that I'd ever sat, I had the same preparation anxiety. I dressed up smart, I can remember feeling similar. I took it seriously, going to the toilet too many times and trying not to drink too much coffee. This feeling pre viva was not dissimilar to interviews for medical jobs that I have had. It's the kind of weight of expectation of the unknown versus the weight of what it means to get it or not get it.

My viva was somewhere between two and a half and three hours. There were not any breaks, I think the vivid things that I remember from the viva, one is when I sat down both examiners opposite me, one of them had a copy of my soft-bound thesis, and they had about two hundred little post-it notes sticking out from the edges of all the pages. I saw all these little tabs coming out the book, and my heart kind of sank at this point. He was stroking it, a bit like a James Bond villain. He said 'Yes, we're going to have fun today.' The second main thing that I remember was that it became clear to me that these two people had read my thesis. They had read it in its entirety, and they'd read it and thought about it. It was the most engaging two and a half hours of talking about what I'd been working to the nth degree for kind of three years, with two people that had read it, thought about it, understood it. It was only then that I realised how few other people had read it.

My supervisors had commented on individual bits of my thesis over the years. So, I had never had a single conversation about the whole entirety. Likewise, no one else that you met, like your mum or your wife or your friends, or anyone else that you've ever talked about with your PhD has ever, apart from what you've told them, had ever gone away and read it, and come back and said, 'Oh, I liked that bit'. So here I was with two people who had read it and could talk about it. And actually, that preparation about me being the person in the room who knew the most about the thesis was true, but also, I was talking to two people that I didn't have to explain everything to. Both my examiners were resuscitation science workers, they were clinicians, they knew the topic area, they knew some of the background stuff. One was paediatric, one was more adult, but they were both very fair. And they were interested in what I'd written, they were interested in the ideas that I had. They challenged me on some of the assumptions, but from a position of curiosity rather than criticism. This was the most enjoyable part of the thesis; actually coming out going I've just spent two and a half hours finally being able to talk to someone who actually has fully understood my thesis. My examiners did not go off-piste, and they certainly did not play themselves against each other, or try to show that they were more intelligent, had done more in this area and so it was supportive, it was interesting, it was engaging, and it was quite fun being able to finally talk to someone who had read it and wanted to know more.

I think at the end of my viva they made me stand outside, they talked to each other, and then they invited me in, and then they probably said something quirky or tried to wind me up. Then they said 'Congratulations you've passed' and shook my hand. I don't think I was quite expecting that. I'm not quite sure what I expected at the end, I think, you know, that they would hand me hundreds of corrections, sort it out, or tell if you'd passed at another point. But actually, it was great. I was like, 'wow'. Then I think I ended up going and having lunch with the internal examiner. It was all a bit surreal, and I was like, 'Oh, wow, I've passed'.

I passed with minor corrections, which included some of the clarification of the processes. It is interesting to note that all those little tables I was like, 'Oh, my god, are those all errors or are those all questions'. I had a fascinating inability or one of the examiners' particular pet hates was the use of 'because of' or 'due to', and whatever I did was the complete opposite of what he liked. That just involved a 'control F' to find every time I wrote the words 'due to' I changed a 'because of' or the other way round. I think that 56 of my corrections were just on that, the language that I used. I got the corrections typed up and one of them must have dictated them to his secretary as they were in a page-by-page document.

In terms of it sinking in that I had passed my PhD, it was probably the PhD ceremony. My parents came, and my wife came, and we dressed up, and I had the silly hat, and I did the process. I think it was probably at the graduation that it made it real. I seem to have lots of letters after my name and have done various Bachelor's of Science and Master's of Science, and then my membership exams for the Royal College, and actually for quite a few of them I didn't ever bother going to any of the graduations. I did go to my medical graduation, that was a big one, when I finished medical school, but when I got my Master's I did not bother because that was a really painful process. When I got my membership, you get invited to a dinner at the Royal College of Paediatrics, and again, I think I was working, so I did not go to that. As a result of not attending these events there was a build-up of all these other events. I think it was important, particularly for my wife, she wanted to see the ceremony because I deserved it and she had gone through the pain of it and lived as a partner of a PhD. It was important for them as well as me. I think that was a big day and those two moments were memorable: the handshake at the end of the viva, and then getting the handshake on the table and getting your photo taken.

My positive memory was this realisation of having such an engaging conversation with two examiners who had given time to read, understand, and be curious about my work. That vindicated the whole thing: the effort and the time. That really was the most important thing that came out of it. And I think if anyone sits a viva and doesn't at least get some of that, I think that is a big disappointment.

Reflecting on the PhD experience as a clinician with over twenty years of experience of working clinically in paediatric intensive care, people always say that a PhD is a journey and a lesson in understanding academia and I think any clinician who goes on that PhD journey will have to have gone through a transition from the demands of clinical activity and the demands on their time. The main PhD journey was around discovering the self-control and the difference between what the focusing on academic work is compared to the task-driven daily clinical training and clinical delivery that you need to do at the bedside. The PhD teaches you that, which is a really important lesson that no one really tells you about. I can remember some people told me about that for the first year, and these were old-school consultants who were giving me advice, said that for the first year of their PhD they were banned from doing any clinical work. They were told you have to stop all clinical work because you need to embed yourself in research and just not be distracted by it. Of course, if you're financially tied into needing to deliver both models that is not a luxury, but I had that at the back of my mind.

I think working clinically throughout a PhD reminds you why you are doing what you are doing, and I think it is the same with clinical academics. We wear two big hats. Some people find those hats heavy; some people find them complementary; some people find them a great distraction from one and the other. Over ten years having been through that, I think I've learnt to appreciate the clinical work more because of it. When I put on my clinical hat I see it as a privilege that I'm doing that and very much focus on I am wearing this hat today and to try to stop the multitasking balancing of everything that's going on."

Nursing

Donna Austin, a registered advanced nurse practitioner in a Paediatric Intensive Care unit, completed her *PhD in Nursing* from 2009 to 2020 at University of Southampton, UK.

"I am a paediatric nurse and I completed my PhD alongside working in the NHS. My PhD title was 'Children and young people's experience of source and protective isolation while in hospital'.

Prior to the pandemic, being in single room isolation was a very different experience for children in hospital, compared with those nursed in main ward bays. This study served to understand that experience, to enable care to be tailored to the children's needs. The study used a qualitative, narrative inquiry method to explore the experiences of children in isolation in one regional specialist hospital in the south of England. Data were collected from children (aged 6–17 years), parents and staff from across the multidisciplinary team, using retrospective interviews and video diaries. Data were collected between 2011 and 2015. Data were analysed using a narrative analysis and the three overarching themes that emerged from the data were control, community and coping. Although this study cannot give one structured approach to providing care for children in isolation, it encourages practitioners to consider the children's narratives from this study within their own clinical setting and individualise care according to their needs.

My PhD took a long time! I commenced my PhD in 2009 and began data collection in 2011. I completed data collection in 2015. I had maternity leave between 2016–2017. I completed my first viva in 2019 and I was asked to complete major amendments. I had to re-sit my viva later that year with the major amendments submission, at which point I passed the viva, and I had some minor amendments which were submitted and thereby completed my PhD in 2020. It was a long road, with many hiccups along the way with a challenging group to collect data from and difficulties in methodology.

To help prepare for my viva I did a practice viva prior to my first viva in 2019 with my supervisors. In honesty I didn't do enough preparation to be able to defend my decisions and processes in my thesis. I knew my subject and the data that I had collected, but I did not answer their questions with sufficient conviction to convince them, this led to me doing the rewrite and repeating the viva. I had a mock viva with my supervisors.

Before going in the first time, I felt completely unaware of what was to come. I was not sure what to expect and on reflection, completely naïve to the level of detail that I was going to be expected to give to defend my thesis. When people say defend your thesis, they really mean DEFEND! I was not ready for that. After I came out, I knew that I was going to have a lot of amendments as they had already alluded to that in the viva. I was just reeling though; I don't really know what I thought the outcome might be.

The second time, I felt so prepared! I had done multiple mock vivas with friends, driving in the car I would answer potential questions out loud. I had read and re-read everything that I had written. I felt confident that no matter what they threw at me, I could answer it and I would answer it with conviction. Coming out, I knew that it had gone much better, and I hoped that it would be sufficient to be awarded the PhD (although I was still very surprised when they said it was sufficient!).

In both the vivas I was privileged to have an internal examiner who was an adult nurse by background, and she had researched widely within the nursing and qualitative research field. The other examiner was an external from another university who was a paediatric nurse by background and she knew the world of paediatric research and family-centred care very well. My lead supervisor sat in on the viva, only to observe.

Both of my vivas were between 1.5 and 2 h. I can't recall exactly how long! I didn't feel the need to take any breaks during the viva.

In the first viva, I was surprised with the questions asking my opinion. Despite living and breathing this study for so many years, I was surprised to hear that the examiners wanted to hear me argue for my opinions on seminal papers within my field of nursing and be critical of them. I did not feel that I had the credibility to be critical of researchers that had been in the field for years, regardless of what my data showed. This was one of the key pieces of advice from the examiners for my second viva.

I did not enjoy my first viva at all. The second one, I knew what to expect a lot more and I was much better prepared to argue. I think in hindsight a greater level of preparation to defend my research would have been helped me a lot further.

The result of my first viva were major corrections and resit of viva. In the second viva I received minor corrections. I was told at both vivas what the amendments needed to be. These were backed up with an email to ensure that I had understood all the details. For the major amendments, I was given six months. For the minor amendments, I was given six months (it would have been three months; however, this was just as the COVID-19 pandemic had started and I was required to work full-time hours clinically as a nurse in paediatric intensive care, so with this knowledge the examiners suggested six months).

The major amendments from my first viva included greater explanation in the text of the journey that I had been on. I had to resubmit to ethics four times for various amendments due to minimal recruitment of children to the study. I was really passionate that the child's voice should be central to this study and as such I had to adapt the research methods a number of times to ensure that they were heard in this study. However, the clarity and detail of these adaptations did not come across in the first submission.

Another area that required major amendments was the explanation of the analysis process. The examiners stated that in the viva I was able to articulate the analysis process well, but this hadn't come through with sufficient clarity in the thesis. The final area that they wanted major rewriting was the implications for this research. Again, at viva, the examiners felt I could articulate the final chapter much better than it had been written.

In honesty, this was the last thing that I had written within my PhD thesis and with a close deadline (and a newborn baby), I probably did not give this section sufficient attention and detail. As a clinical practitioner, this is the section of the thesis that is most relevant, so as soon as they had highlighted that there was a dearth of suggestions for practice, research and education, I could easily amend this area with more detail.

In the second viva, my amendments were mainly typos or layout errors. There were a couple of sections which required an additional sentence to link or explain something in more detail, but these amendments were made in hours, compared with the significant rewrite of the major amendments.

The best thing from my viva that I can remember is being challenged on something you know inside and out and being able to give answer with the literature to back it up, but also your own primary research data. For me, the most positive thing was also using the children's voices (my participants) as data, to be their advocate, and share their experiences in an area that they would not normally be heard."

Emma Wadey, a registered nurse, completed her *PhD in Nursing* in 2022 in the UK.

"My PhD study aimed to understand the impact of stigma on how parents cope with grief after the suicide of an adult child. A systematic literature review informed the design of a longitudinal study that used a constructivist-grounded theory approach. In total, 12 parents from south-east England, of adult children who had died by suicide, were recruited. Semi-structured interviews were conducted at two time points between 2015 and 2016.

I needed to take several significant breaks during my course of research due to the following reasons, broken arm, emotional burden of the research subject and then finally a year off due to my national clinical role during the pandemic. Therefore, the PhD took officially 54 months part-time but was over an eight-year period.

To prepare for my viva I attended a mandatory workshop on PhD Viva preparation run by the Doctoral college at my university. This was really helpful and informative in understanding what to expect and how to prepare. I organised a mock viva to take place after final submission but six weeks before the viva exam. This was led by the head of the Doctoral college and an expert in qualitative research methodology. I prepared thoroughly, using a list of likely questions and the feedback from the mock viva; however, some of the questions differed considerably from these.

I was much more nervous than expected going into the viva regardless of the preparation. This was mainly because it was the first time that an expert in the field of study would read my research. Afterwards, I felt overwhelmed when hearing the feedback and it was difficult to focus on the positives. Although there was huge relief because there was more work to do and a need to resubmit, I didn't feel able to celebrate.

In my viva exam I had a chair present as both examiners did not meet the Doctoral college requirements on enough prior experience of examining PhDs. The internal examiner was an expert in qualitative methodology. The external examiner I had identified as they were an expert in the field of suicide research. In total, the viva, which was conducted virtually, lasted approximately 90 min. The examiners requested a short 10-minute break halfway through. In the viva there was a focus on the content of the quotes used, with the feedback that these were distressing and were required to be shortened and some suicide details to be removed. There was also a focus on the layout of some chapters rather than the content.

Once I had overcome my nerves, I had enjoyed the mock viva as an opportunity to talk about my research. This was not the case within the viva exam. I found it very stressful. I experienced it as more critical than constructive, with less opportunity to discuss the findings and strengths of the research. Interestingly, the feedback I had from the examiners was counter to this as they experienced the viva as engaging, interesting and me as articulate and passionate about the subject!

My outcome was that I had successfully passed the viva exam but had minor revisions to make before resubmission. I received the conditions for corrections ten working days after the viva exam. I was given six months to complete them but actually successfully resubmitted six weeks later. The conditions mainly including reorganising some sections; for instance, moving the theories of grief and stigma chapter to before the systematic review and combining the two findings' chapters into one. I was asked to move the table of included studies from the appendix to the main text and to move another table to an appendix. I was also asked to reduce the length of quotes used and remove all details about the methods of suicide.

On successful completion of the viva, there was a sense of achievement that such a significant milestone had been met. Meeting with my supervisors directly afterwards was probably the highlight, as although at that point I was still a bit overwhelmed it was joyous to see how pleased and proud they were."

Adrienne Hudson, a *registered nurse,* completed her PhD in Life and Health Sciences in the UK in 2014. She now works as a Nurse Researcher in a tertiary children's hospital in Australia. As a Nurse Researcher, Adrienne's role is varied. She conducts her own research and supports other nurses, doctors, and allied health to refine their research ideas and progress their research. Her role is about building the capacity and capability of clinical staff to undertake research. Skills that she developed during her PhD, like working with teams, leadership, managing budgets and timelines, and applying different methodologies or approaches to solving problems, have also enabled Adrienne to work in different clinical roles within the hospital, like Nursing Director roles. Many academics in nursing move across to working in universities but completing a PhD and the skills acquired along the journey opens many opportunities in the acute care settings for nurses. Adrienne shares her experiences:

"I started my PhD in April 2010. I was lucky enough to secure a scholarship from the local health authority which paid my existing wages, university fees and study costs. The aim of the scholarship was to support clinicians to remain in the hospital system and choose clinically relevant topics that would enable them to remain in the acute care setting post PhD. Through my PhD I wanted to explore what the experience of caring for a child who has an acute life-threatening event (ALTE) is like for healthcare workers in a paediatric hospital. I then used this information as part of a mixed-methods study to develop interventions aimed at preparing and supporting staff for the potential psychological impact of caring for a child who has an ALTE.

It took me three years to complete my PhD. I started in April 2010 and handed in my thesis in November 2013. I had my first child at the end of my first year so I took six months' maternity leave during my PhD. I was 39 weeks pregnant during my viva with my second child!

I used a mixed-methods approach to prepare for my viva! I had never done one before so I didn't know what to expect which makes it hard to prepare. I had several sessions with my supervisors to practice. The first session was more an information session where my supervisors clarified things like how long a viva might be, the format of the session (examiners would be present with a chairperson), the rough timing of the viva and the practicalities of what I might expect. Subsequent sessions were practice sessions where my supervisors would ask questions they thought I might be asked or I would put forward things I thought I might be asked and then I would practice answering them. They then gave me feedback to help me refine my answers. Even the *simple* question of "summarise what your PhD is about" took quite a bit of practice to get a coherent, succinct answer, even though I had lived and breathed this work for three years.

To prepare for the viva I also met with colleagues from different disciplines who had done their viva. They were in different fields so theirs were a bit different in format, but it was handy to understand some of the emotions and timings I might expect. I also read through my thesis and the data I had collected several times before my viva because it had been quite a few weeks (I want to say four to six weeks) between handing it in and my actual viva date.

Leading up to my viva I knew I was preparing for one of the most important 'interviews' of my life. I like to be thoroughly prepared for things and well-practiced when I am presenting at conferences and the like. So, it was hard to prepare when I didn't really know what I was going to be asked. I did get to a point where I could rationalise that most of the questions I was going to be asked were about this programme of work that I had invested three years in doing and if anyone knew this work from start to finish it was me. When I was able to frame it that way, I felt more prepared and mentally ready for the challenge.

My supervisors suggested prior to the viva to put tabs in the thesis so that I could quickly locate important sections without having to fumble around on the day. So I put coloured tabs on each chapter so I could locate those quickly. I also put different coloured tabs in pertinent parts of the chapters (results, for example) so I could locate them quickly. This really helped during the viva because there were some chapters that we skipped over quickly (or altogether), and it was one less thing to try and juggle or be distracted by in the middle of the viva.

On the day of the viva my mum came with me to the university and waited in the office for me. She had travelled from Australia for the birth of the new baby, but I was so happy she could be part of this experience with me. Two of my supervisors came and sat with me as I waited to be called for the viva. They chatted and kept me company. When the administrative assistant came to get me, one of my supervisors walked with me and chatted with me to keep me calm. My viva was 4 p.m. on a Friday afternoon. It only took about an hour and forty minutes, which is apparently considered a short length of time. When I met my examiners in the viva, I clearly remember the external examiner saying to me 'You are very pregnant, and I have a train to catch so we will be out of here by 6 p.m. at the latest'. It was an ice breaker and gave me a good idea of how long things would take. In the room with me was the chair, an internal reviewer (who was external to my supervision team) and an external reviewer. I was offered a break but was happy to keep going and did not need a break.

I don't remember being particularly surprised by the questions. The external examiner was considered one of the lead experts in the theoretical framework I used. She focused on the theoretical framework and did not have much to add to the other aspects of the work. Her approach was far more conversational and genuine interest in the work that I had done. She had a few questions to clarify the clinical aspects of my work, but the main points from her were tips on things to consider when I published the work. It felt very supportive and positive.

I was a little surprised that the internal examiner had comments and feedback about the theoretical framework given the expertise of the external examiner. The internal examiner had an eye for detail so she picked up a lot of the grammar and punctuation corrections in the rest of the thesis. She went through chapter by chapter with questions and feedback. At the end of the session she handed me her copy of the thesis with notes so that I could use it as a guide for my corrections, which was really helpful.

I remember walking out of the viva and feeling a bit shell-shocked. I wasn't 100% sure I had passed and had to check with my supervisor (who had been waiting patiently outside) if I had actually passed on the walk back to the office. My supervisors asked me what questions they asked and it was hard to remember them. That was ironic as part of my PhD talks about the forgetfulness when you are in a stressful situation due to the effect of adrenalin on your body. That shell-shocked feeling was also because I had lived and breathed this PhD for three years. I always knew there was a viva at the end and had done the preparation and then it felt like it was over so quickly. I kept waiting for someone to tell me they had made a mistake and I hadn't really passed and that feeling lasted for many months. On the other hand, I was so proud that I had actually done it. I had done a PhD and passed with minor corrections. It felt like one of those experiences that only someone else who had done their PhD could ever understand. The viva was a culmination of the entire journey and I had done it!!!

I passed my viva with minor corrections. One of my corrections was regarding the methodology used. I used interpretative phenomenological analysis (IPA) as one of my methodologies. As part of the methodology the researcher needs to be aware of not imposing their own beliefs/frames when interpreting the participant account. I distinctly remember my external examiner talking me through the difference between separating the researchers' beliefs and 'bridling' them. So not to try and separate beliefs (it's almost impossible to do this) but to try and be aware of them and how they might influence your interpretation. This involved a reflective process, and it has always stuck with me.

I really loved my entire PhD journey. I recognise I was very fortunate that I was awarded the scholarship that paid my wages, fees, and research costs. I was also fortunate that the purpose of the scholarship was to keep nurses who were doing a PhD in the hospital system. I tell people it was my job to do my PhD but I really enjoyed the journey and the experience. The skills I developed during my PhD have translated into transferrable skills that have afforded me opportunities outside of research. As mentioned earlier, I have led some big projects across the organisation and acted in Nursing Director positions. I also had an amazingly supportive supervisory team. When I was first applying for the scholarship, someone mentioned one of my supervisors and suggested I contact her and ask if she could help me refine my research question and methodology because she had experience in health services research. We arranged to meet—I took my clinical supervisor with me and she brought along the person who ultimately became my primary supervisor. They spent three hours with me helping me to refine my research question and planning the methodology to use to answer my question. I knew then and there that if they were willing to invest that much time and effort into me with no guarantee of money they would be the best support for me if I did embark on the journey. I wasn't wrong! I always tell people that supervision is absolutely key. You need good supervisors that will support you through your journey. I did my degree at a university that did not even teach my profession. That didn't matter because I had the right supervisory team."

Theology

Ian Paul completed his *PhD in Theology* in 1999. Ian now writes a very popular and well-read blog and various books. Ian shares his experience of the viva:

"My PhD was a mix of biblical studies and philosophy of language, looking at the interpretation of metaphorical language in the Book of Revelation, drawing on the work of Paul Ricoeur. I completed my PhD in six years studying part-time.

To prepare for my viva I re-read the thesis and familiarised myself with the literature. I did not have a mock viva. In my viva the following people were present: external and internal examiners; and my supervisor, who simply observed. It lasted one and a half hours. I did not take any breaks during the viva.

No, I was not surprised by any of the questions. The toughest question was probing the single most demanded resource I had explored, setting out Russian formalist and structuralist approaches to language which influenced Ricoeur's own thinking. I can still remember the main argument of this article to the present day; my supervisor, at tea afterwards, gasped at it and said 'I don't think I could have answered that'! I afterwards also learnt that the internal was very helpful, having deflected an unreasonable question the external wanted to ask.

With regards to enjoying the viva—yes, and no. It was a real, testing, examination, but the outcome was positive, and because I performed well it felt like an achievement. I passed with minor corrections all of which were typos—nothing of substance. I was given three months to complete these.

The viva was an intense experience. But I had worked in industry and business previously, and taking a viva is nothing compared with being responsible for the lives of people on a production line!

One of my positive memories of my viva was that my supervisor was wonderfully supportive, and we had tea in the Queen's Lane Coffee House afterwards, which he paid for! He afterwards told me he had been in no doubt that I would pass."

Education

Linda Golding completed her *Doctorate in Education* 2014, from a university in the UK. Here Linda shares her experience of preparing for the viva and the viva itself:

"My Doctorate was in education management and leadership, and I completed this in 2014 in the UK. It was a case study about leadership and management projects in Further & Higher Education and their impact on practice titled 'Planning and managing change projects in a college of FE/HE'. The experiences and perspectives of one team were explored with ideas as to how future practice could be improved.

I was studying part-time alongside a demanding full-time job in education, so it took eight elapsed years in total with two one-year breaks during this time. The prerequisites were to have achieved a Master's-level degree (which took four years) within the last two years and to have achieved at least 70% on the Methods & Methodology modules and at least 70% in each of the two Education Management and Leadership modules prior to starting the doctorate.

I prepared for my viva thoroughly in the following ways and at the following time points. Immediately prior to viva:

- I had a good night's sleep.
- I stayed close to the viva venue the night before to avoid travel complications.
- I wore my favourite outfit, so as to feel good about myself.
- I arrived at the venue 60 min early so I could familiarize myself with the surroundings, refreshment, and washroom facilities. Also, so I could just sit and relax and gather my thoughts in a calm space for 45 min prior to the viva.

I spent 15 full days preparing for my viva and took two weeks paid time off work in order to do this. In fact, almost all of my holiday entitlement from my full-time job was taken as study leave, including most weekends for the duration of my doctorate studies. This had a significant negative impact on my social life.

- I attended a two-day workshop 'In Preparation for your Viva', which included sessions in which each attendee spent 30 min presenting their own thesis and then 30 min answering questions from the attendees and the academics running the workshop.
- I read, made notes and identified potential questions from a useful book entitled *How to survive your Viva: defending a thesis in an oral examination* by Rowena Murray.
- I produced a coloured A3 page summary of my whole thesis and how all the parts joined together using diagrams with arrows and bullet points so I

could easily see it and refer to it during the viva. It provided a useful summary and prompt to help me remember key names and points about which I was worried I may forget under the pressure of an oral exam.

- I annotated, highlighted key points, and attached post-it notes to key pages of my copy of the thesis which I took to the viva. This enabled me to easily identify points in my thesis should I need to refer to them in the viva. My thinking was that although the examiners were experts and specialists in related fields, they may have only skim-read my thesis; therefore I was the expert on my own thesis and the reasoning behind my decisions.
- I practised many times out loud, taking time to answer slowly and clearly potential questions which I had devised using a variety of sources and I treated the viva like an oral presentation and a defence of my thesis. For example, Why this sponsor?; Why this topic?; What was my relationship to the sponsors and the participants and how did this influence my research?; Why these participants?; Why this methodology?; Why these theories and models?; How did I analyse my data?; How did I arrive at my findings and conclusions?; and What was my contribution to the relevant body of knowledge? I also rehearsed my views about the strengths and weaknesses of my thesis and what I could have done differently and why.
- Whilst doing the detailed preparation for the viva I found a few mistakes in my thesis. I typed these and took the list with me to the viva.

Additionally, I also had a mock viva with my research supervisor and two of her colleagues. It was two hours and as realistic as they could make it, including using a room at a venue in London, i.e. unfamiliar surroundings. I had a one-hour de-brief with my supervisor after this mock viva in which she gave me pointers as to what I did well and what I could have done better and asked for my opinion on how I thought it went. This mock was invaluable to the learning experience and boosted my confidence. During it, I learnt to give in gracefully if the examiners made a point about something I had not previously considered, and I also learnt that it was OK to say 'I hadn't considered that' or 'Can you repeat the question please?' I also practised listening to the question, giving considered responses and maintaining eye contact with all the examiners and not just the one asking the particular question.

Before the viva I was extremely nervous and worried that I might forget key theorists and the literature. I was also apprehensive and worried that I would not be able to speak fluently without drying up or stuttering. I was also lacking in confidence that I would be able to hold my own and answer questions from a theoretical viewpoint and an appropriate academic level, especially if unexpected questions were posed about areas in which I had no knowledge.

In my viva, my research supervisor was present but as an observer only; she sat at the back of the room and was not allowed to contribute or comment in any way. Two external examiners (one of whom chaired the viva), as specialists in the area of education management and leadership in Further Education or Higher Education. They were both from a different university to the one where I was studying and there were also two academic doctors from the university where I was studying. I had never met any of these four panel members previously. I was given the panel members' names two weeks before the viva so I researched their areas of expertise and research papers so as to be aware of their interests and potential viewpoints on my subject area.

My viva took three hours excluding the 30-minute recess—we overran the planned two hours because we got into some interesting and constructive debates and there were several questions from the examiners trying to elicit my thinking behind relevant points and to clarify and check my understanding about points in the thesis which they felt were not sufficiently clear. There were no breaks, except for 30 min at the end of the viva when the panel retired to discuss and agree on the viva result—and their feedback took another 10 min. Any other breaks would have been disruptive to thought processes—water was provided.

In the viva, 50% of questions were as anticipated and had been mentioned in the viva book or workshop, including some relating to my original reasoning and thinking, which I had long forgotten (prior to my preparation). But 50% of the questions were specific to my thesis, checking my definitions and understanding about particular points which the external examiners felt I had not clarified or explained in sufficient depth.

I absolutely enjoyed my viva, it was my opportunity to orally present my research and findings after many years of work. The examiners were clear and fair in their questioning, mostly asking me to explain my reasoning and thoughts about points in the thesis but also asking, for example: 'As a researcher-participant, what was your relationship with the sponsors and the participants and how did this affect your research and findings?'; 'Had I considered x... ?'; 'Why did I not include x... ?'; 'Can you explain what you meant by x?'; 'How did you analyse your data... and why not use x as an analysis method?'; 'How did you reach your conclusions?' And 'What alternatives could you have used for some of the above points?'

Immediately, afterwards, I felt extremely positive and highly motivated, elated and relieved and very surprised that I actually enjoyed the challenge which the viva presented. I was really pleased with myself that I thought I had taken my time thinking about the answers to each question before speaking and even managing to make notes during the viva to help stem my nerves. I felt, quite surprisingly, that this viva event was a good learning experience in itself.

I passed with minor amendments with a three-month deadline to resubmit from the date that the email was sent by the external examiner. These amendments I received verbally, after a 30-minute break at the end of the viva when the panel retired to discuss their opinions prior to returning to give me their verdict. I was making detailed notes throughout the viva, which were invaluable in making the minor amendments to reflect the examiners' concerns. I also received details of the required amendments (half an A4 sheet) in writing by email sent by the external examiner chair of the viva within three working days of the actual viva. This included an automated 'Read Request' confirmation as well as a request to confirm by email that I had received and understood the amendments required. In addition, on the next day, I received a written copy by post to my home address. Some of my corrections included:

- Add two paragraphs to Chapter *n* about my understanding of shared leadership and how I was interpreting leadership models in this study.
- Ensure you refer to the composite model (a combination of theoretical models which I had chosen to explore) appropriately and consistently throughout this thesis.
- Summarise the concepts of validity that you use and explain their meanings. Then relate their application throughout the Methodology chapter back to this summary.
- Revise and develop Section *n.n* to show the key contributions that this study has made to the knowledge base in this area (i.e., beyond the specific case) and relate them briefly to existing literature and theories. In the viva, the examiners mentioned several key contributions of my thesis which I had not identified.
- Check the whole thesis very carefully for grammar, punctuation, syntax and expression.

I have a strong recollection of feeling positive emotions at the end of the viva when the panel gave their verdict and said that: 'We are confident and all agree that you have achieved the required standard for a doctorate in education but we need you to do some minor amendments to your thesis… you have satisfied us that you understand these points but have not clarified them explicitly or sufficiently in your written thesis… you have three months to complete these amendments… and you must meet this deadline if this Thesis is to be given an achievement award.' This, for me, was a particularly positive viva memory."

Pharmacy

Julie Menzies completed her *PhD in Pharmacy and Nursing*.

"I undertook my PhD within the School of Pharmacy, University of Birmingham, UK, 2010–2018. I'm a Registered Nurse by background but had a special interest in the design of clinical trials and therefore developed a PhD proposal with a Professor of Pharmacology and two Consultant Intensivists.

It took eight years in total from start to submission of my thesis. I'm not sure I would have started if I'd realised how long it was going to take! I was fortunate to have been awarded a fully funded PhD scholarship which enabled me to continue in my NHS role and conduct the PhD part-time. I was due to undertake it over four years but extended it with two periods of maternity leave and a subsequent reduction in my hours.

I started preparing for the viva approximately seven weeks before the event. For me this felt the 'right' amount, particularly with no funded time left for preparations and having to do everything in evenings or at weekends. I'm a paper and notes person so for me there was a lot of note making involved in preparing for the viva exam.

To prepare, I started with a page-by-page, line-by-line review of my thesis so I was fully immersed in the content. I reviewed key terms and legislation to ensure I was confident in definitions and governance. I reviewed key references and made notes on their methods and results to ensure I could critique them if required. I read extensively about philosophical considerations because I knew this was an area where I could struggle to communicate clearly and summarised key phrases to help in the viva. I made an A4 mind map of each chapter to help me pin down key content in each chapter. I followed an approach recommended by Nathan Ryder (2016) in *The Viva Prep Handbook* called VIVA: Valuable (to others), Interesting (to you), Vague (or unclear) and Ask (questions). I conducted this for each chapter, and this forced me to be objective about where there was a lack of clarity, what my key messages were and to consider what questions might arise. I also spent days preparing responses to common viva questions. Although I started off worried that there were so many questions that could be asked, I came to realise that there are a number of common areas which could be prepared.

I had two separate meetings with my PhD supervisors approximately two weeks before my viva. These were useful to review and prepare responses on the areas where we knew I was not feeling as confident. I also had a mock viva with someone external to my supervisory team and university, which I would highly recommend. Although at the time it was stressful, it proved very useful to have a fresh set of eyes. Their questions gave me some clear areas to review and consider better ways to articulate my decision-making.

I remember in the week leading up to the viva feeling absolutely terrified; sick with anxiety, extremely snappy and not sleeping. I felt such a huge sense of responsibility to complete this last step with the PhD, not only for me but also for my husband and children. I finally felt at peace around 9 p.m. the night before the viva. I finished reviewing my notes and the realisation came that I had done everything I could to prepare, and I was ready to 'defend' my thesis. The nerves didn't go completely, but from then on I felt much calmer.

In my viva, there was an internal and external examiner and a chairperson. The viva lasted just under three hours. I remember feeling surprised that that they seemed to be wrapping up as it had flown by. I do not remember taking any breaks. The viva felt well-paced and there were refreshments available, so I didn't feel I needed a break.

I was pleasantly surprised by some of the questions! The questions seemed reasonable and appropriate, and I was able to answer them all. The key element for me was realising that they were genuinely interested in my responses; not to catch me out, but to hear my justifications and engage in academic discussion.

I hadn't expected to enjoy the viva at all, but the viva was a really lovely experience to be able to talk in depth about the work, my decision-making, the challenges I'd encountered and the interpretations I had spent months agonising over. I was so passionate about my PhD work all the way through, but by the end of writing the thesis I was absolutely sick of it. So was everyone else around me, from my supervisors to my family and friends! The viva was therefore the first time in a long time that people seemed genuinely enthused to hear about my work and I really enjoyed being able to talk in depth with people who had read the work and valued my opinion.

I was lucky that I didn't have to wait long after my viva finished for them to tell me I had passed with minor corrections. It honestly felt like a huge weight had been lifted and I sobbed immediately. There are so many hard moments during a PhD, but this moment really made it all worthwhile. I received the corrections as a hard copy of my thesis with post-it notes stuck on relevant pages. I also received an email summary of all the key corrections required. I completed all of them within eight weeks, which was well within the time I was allowed. Some examples included:

- 'The abstract does not do justice to the work… there is a lot of interesting data within the thesis that is not captured within the abstract.' The abstract was only 200 words, so it was difficult to add in much detail about the results. However, I did amend this in light of this feedback.
- 'Please be consistent with reporting of school years. Use Year 12 and 13.'
- 'Headings and legend need to be on each page to assist reader with a large table.'
- 'Some of the titles on figures and tables are very small. Please can you increase in size?'

The corrections were nearly all of this nature; more about clarity than about addressing flaws.

Finally, my most positive memory during the viva was when the chair told me at the end that he had really enjoyed listening to me and the external examiner (a professor) discussing the different ways that a piece of work could have been conducted. He said it had felt like two panellists of equal calibre at a conference engaged in academic discussion and debate. This was a huge boost to my confidence and a really proud moment."

4.4   Examiner Experiences

Dr. Juliet Wakefield is a senior lecturer at a university in central England and here Juliet shares her experience of examining vivas:

"I have examined five (four, plus one where I only wrote a report on the thesis and there was no viva, as per the institution's procedure) vivas. These have all been in the role of external examiner.

All the PhDs I have examined have involved research that explores *group-related processes*, and all have included at least some element of the application of the Social Identity Approach in order to understand group phenomena: this is my area of expertise.

I enjoy being able to tell someone that they are now a doctor (pending minor revisions) is a very memorable and special experience. More generally, I enjoy being able to talk to and hear about candidates' research and being able to give them the space and time to talk and think about their work deeply.

I have awarded minor corrections for all five theses I have examined—unless something has gone quite wrong with the thesis, this is by far the most common outcome.

The main thing I look for in a thesis is that there must be clear evidence that the candidate's work has contributed to knowledge and advanced our understanding in some way: this needs to be clearly communicated in the thesis. Also, if the thesis contains multiple studies, it needs to be clear in the narrative of the thesis how they link together and build on each other (rather than being a list of disparate and isolated studies—this is particularly common in theses that consist of multiple papers that have already been submitted to journals). The thesis should also show evidence of the candidate having reflected deeply on various potential interpretations of their findings, as well as on the strengths, limitations, and potential practical applications of the research.

In the viva I am looking for answers that clearly indicate that the candidate wrote the thesis themselves and understands the research they have conducted, as well as how the work contributes to knowledge. Answers that indicate that the candidate is willing to think deeply about the potential strengths and weaknesses of the research they have conducted. Evidence that the candidate is willing to think deeply about the potential interpretations and applications of the research, as well as having a willingness to consider alternative perspectives on the potential interpretations and implications of the research findings. Evidence that the candidate knows what the next steps would be in terms of future research directions.

My top tips for preparing for the viva:

- Don't start preparing too early—two weeks before the viva should be plenty of time.
- Have a hard copy of your thesis which you will have next to you in the viva and to which you can add page-markers so that you can access any part of the thesis within seconds.
- Work through the thesis from start to finish, making notes of any issues or potential questions that you think you could get in the viva.
- Ensure that you do a mock viva—ideally with an academic who knows your work well but is not your main supervisor (your main supervisor may be so close to your thesis that they have blind spots regarding things like potential limitations of the research).
- Take time to relax the day before the viva—I went for a massage before mine!
- In the viva, don't be too quick to blurt out an answer to every question: taking a few seconds to find the relevant page in the thesis gives you valuable thinking time, and it is absolutely fine to ask for the question to be repeated in order to make sure you have understood it correctly.

Every viva is special in its own way, but I did find examining a thesis in Australia (via videocall from the UK) to be especially memorable, because it was so strange to think that I was congratulating the candidate on a successful viva at 11 a.m., but that it was 9 p.m. for her!"

Lyvonne Tume is a professor of critical care nursing at a university in the north-west of the UK.

"I have examined in my career to date three PhDs as external examiner, including one European PhD and two MScs by research thesis as external examiner. These PhDs have not all been in my specific research field, but in my broader research domain.

In a viva I am examining whether the candidate can clearly articulate their ideas clearly, showing this work is their own. When I examine a thesis at doctoral level I am looking for a clear, coherent structure, clear rationale for the study, relevant and critical literature review, clear methods chapter in detail, clearly presented results chapter/s (with no discussion in it), a detailed discussion chapter showing synthesis and new knowledge with a clear identification of the answers to the research questions asked/the aim and objectives of the study and recommendations for future research and for practice. I also (in most theses) like to see and understand the students' chosen theoretical framework underpinning the study. I also ask them what they would do differently if they did this again, to demonstrate their learning through the process of a PhD.

I like the viva to feel like an academic discussion around a piece of work, but the role of the independent chair is very important in a PhD viva to ensure fairness, candidate, and examiner comfort—for example, breaks if needed.

From the PhDs I have examined the most common outcome is major amendments.

My tips for candidates preparing for the viva:

• Research the backgrounds of the examiners and their methods used.

On the internet there are lots of example questions asked by candidates in PhD vivas shared, read these and try and answer these yourself."

Professor Bill Deakin is a neuropsychiatrist and has worked in the field of *neurobiology of mental illness* for the past forty years. Bill shares his experiences of examining PhD theses and conducting vivas here.

"I examined about two theses per year or so. I have been examining over 40 years with fewer lately and initially I would say I have examined around 60 theses in total.

I have mostly been the external examiner. As the internal it feels right to defer somewhat to the external examiner and let them lead. Crucially important to meet beforehand re strengths and weaknesses and to plan the exam. Examiners vary in how much they want to ease tension in the student but there should be an agreed approach from the outset. I have been thrown once or twice by examiners who reassure the student immediately that everything is basically OK and we just want an interesting scientific discussion. I find this may be rather anti-climactic for the student—it is anticipated as a rite of passage with some stress and the examination ought to be rigorous as well as praising good aspects. Furthermore, one cannot know at the outset whether the student has serious gaps of knowledge, fundamental misunderstandings or been over-reliant on results from collaborators. Such problems are very rare, however, and I can't recall outright fails and only a few re-writes. So I am friendly and smile and say the thesis was very interesting to read and represented a lot of work while retaining a slight inscrutability. I like to hear how the student got into the position of carrying out the research, which eases tension, and then ask for a brief three-minute overview and off we go.

In my early years as an examiner, theses were typically in a book form with mostly unpublished results. This required and justified a critical approach to the exam. Nowadays, theses in neuropsychopharmacology may consist of, or include publications which have been refereed. This makes it less likely that there is a major problem with the work and results in a more wide-ranging discussion. Presenting formatted manuscripts has the ethical benefit of increased dissemination of work funded by taxpayers and charities, the collaborative effort of human participants and the sacrifice of animals.

In Scandinavian countries, PhD examination is ceremonial and public. In one, we were dressed up in mediaeval robes and warned not to ask probing questions in front of a be-gowned university assembly and an audience of relatives, friends and well-wishers. In another I had to speak about my work, then lightly interview the student seated in comfortable armchairs in front of an audience, receive a small gift from the student and, with minimal prior notice on the day, give one in return. And then join a big dinner put on by the family with the request 'Would you mind saying a few words afterwards?' This was a long day's work but with good science and an enjoyable social occasion with new people.

I have examined studies using laboratory animals and clinical groups. Always relevant to psychiatric illness. I quite often have areas of ignorance, and this allows the student to help out and show communication skills.

I enjoy getting to grips with a study with the person who ran it, in a viva.

Examiners bend over backwards not to require a rewrite. The most common outcome is minor revisions. Maybe in 10–15% I want to see the thesis again.

Two main things I look for in a thesis are clarity of purpose and clarity of exposition. So first, an introduction that leads to a compelling rationale for the experiments ending in a theoretical/mechanistic statement or hypothesis about what the student thinks may be going on. This leading to a set of tests or predictions from the theory/hypothesis. It bugs me a little if a set of predictions are headed 'Hypotheses'; 'administering drug X will counteract the microglial response to LPS' is a prediction not a hypothesis—there is no mechanism. The discussion should address the questions or hypotheses raised in the introduction.

Clarity of exposition comes from consideration for those who have to read the thesis. So, brevity, logical organisation and a strong and consistent formatting. General rules: no sentence longer than three lines, four absolute maximum; no paragraph longer than 15 lines, half page abs. max.; two-plus subheadings per page; clarity of the subject or inference of each paragraph; avoidance of abbreviations, especially unfamiliar ones; plain English and avoidance of jargon.

In a viva I am looking for clarity of the objectives and ability to summarise the thesis succinctly. Good knowledge of the literature. Implications for future studies and wider implications.

My top tip for PhD students preparing for the viva: Compose a rote summary of the rationale and main findings by speaking it aloud in language an educated person would understand. Practise it aloud. It is a near universal opening question, yet I have seen students thrown by it."

A senior lecturer in *Chemical Sciences* at a university in the UK shares his experiences of examining over 25 PhDs:

"I have examined over 25 PhDs from a range of institutions (External > 15, Internal 10). There are differences between being an external examiner and internal examiner. As external, you have responsibility for the overall quality assurance (QA) of the examination and the award of the degree, benchmarked to other UK institutions via your experience. The role of internal is more of referee, plus QA within the same institution/department/faculty.

As an internal examiner, you have responsibility to arrange the viva (time, date, place) and liaise with the external and candidate. Often you would briefly speak to the candidate's supervisor to check if there are any relevant issues that the examiners should be aware of that have impacted on the candidate's ability to perform the work (e.g. Leaves of Absence, key equipment being broken or access to facilities being denied, etc.). Liaising with the external is important before the viva, to get an indication of what they think of the quality and if they foresee any significant issues. On the day of the viva, the examiners will meet for 20–30 mins before the start of the viva to compare notes and discuss how they want to run the viva (sequentially through the thesis, or topic by topic, roughly how long etc.).

When you are an external examiner, there is an expectation on you to lead the viva since technically it is your opinion on the thesis/viva (that is, if it's good enough to pass) that is the most important. When you are an external it is also more enjoyable because you are more in control of the main discussions and type of questions, although some internals can be extremely passive and expect you to do everything. The experience of being internal often depends on your rapport with the external—if they are a bit of an egotist, then it's going to be a long day trying to keep them under control.

As an external, I've examined mainly in my area of expertise. As an internal the topics have been much broader since the department that I work in is not traditionally the one that my expertise would be classified.

The aspect of examining a PhD that I most enjoy is sitting down with a candidate for two to three hours and discussing (and learning about) their science. With the weak candidates, it's about probing their understanding to hopefully establish that they at least understand the basics of what they've done and try to let them see how they can improve things; however, occasionally it can resemble a tutorial (!). With the stronger candidates it's about trying to challenge them to go further. There is always something new I can learn as an examiner from every viva.

The outcome seems to (weakly) vary on the quality of the institution from my experience in that mostly the outcome is determined by three things, the effort, commitment and quality of the candidate. The higher the ranking of an institution/department the more those qualities seem to have been sifted by the process of appointing the students in the first place.

I've found that usually most PhDs I examine the outcome is minor corrections, but there are occasionally major ones. Major corrections have usually arisen from missing fundamental issues that might impact the interpretation of results—in turn that typically seems to occur when: (i) the topic is a bit off-centre from the supervisor's main interests; (ii) the candidate has decided (for whatever reason) not to listen/follow advice from the supervisor; or (iii) the candidate has not read around their research topic.

Another crucial point for candidates to realise is that the boundary between minor and major corrections can change from institution to institution—mostly it is about the extent and importance of the corrections needed, but sometimes it can even be that the choice will depend on how long the examiners think it'll take the candidate to make the changes. There are usually regulations determining whether any further corrections would be allowed to minor/major corrections, so this can come into the decision too. From the candidate's point of view, the major/minor boundary is a matter of pride or ego, and occasionally they will try to negotiate that the paperwork reflects 'minors' but essentially it doesn't matter so long as the thesis gets corrected appropriately. Recently, university departments have started reporting the numbers of majors and minors, purportedly as a marker of quality of supervision/research, and so there can be institutional pressure on the examiners to award minor corrections.

In a thesis there are several aspects that I look for and these are presented here in no particular order:

1. Length. Nobody wants to read 400 pages, and it's quite common for overly long theses to be rejected by an examiner so best not let your *magnum opus* run to more than one volume. Similarly, if the thesis is under 150 pages, it'd better be *really* good. Length is a tricky one, because we use it as a rough indication of effort/diligence, but clearly a well written 150-page thesis is going to do better than 250 pages full of padding and waffle.

2. The results chapters. How many are there? Is it a reasonable amount of work? As a physical scientist, I would normally expect four or five results chapters, but have seen as many as ten from star students and as few as one from (very) weak ones.

3. Have the results been published?—that gives some assurance that peer review will have picked up most glaring errors (although you can't wholly rely on it). In the ideal world, you'll find that each results chapter has been published or submitted for publication.

4. Context, originality, novelty. These must be there by definition and obviously will be met if the results have been published.

5. The basics. Is there a good introduction that covers basic theory and background relevant to the following work? Is there a link to basic science etc. in the discussion sections, or it is just observations? Do these sections 'link forward' to the results and discussion later, or are they full of formulaic material that is never referred to again?

6. Quality of the discussion. How carefully have the results been analysed? Have they been analysed using the appropriate methods/theory and does the candidate understand the limitations of those methods? Have the results been linked well to theory and literature? Essentially, it's about whether the discussion attempts to rationalise the results, explain why, etc., and not just describe them and leave the reader thinking 'so what?'.

7. Who is it written for? A researcher, or grandma? Always a favourite one this… don't write it for granny, so it doesn't have to be an encyclopaedic textbook explaining everything in minute detail. The only people to ever read it again will be the candidate and a handful of others. Everyone else will read the publications derived from it, and granny will thank you for it, put it on the shelf with pride, and quite probably only read the Acknowledgements page to see who's been mentioned in there.

8. Presentation. Is the language style appropriate, and has care been taken over getting the spelling/grammar correct and references/figures/tables numbered in numerical order? With everyone using word-processing software, there is little patience with basic spelling/grammar mistakes… it just looks careless and sloppy. Similarly, the figures—are they big enough to read clearly and do they have explanatory legends/captions? Please do not use Excel—it's useful for analysing data and a first look at graphing, but it is not of publishable quality. Get a student licence for Origin, or SigmaPlot or something similar (other graphing software packages are available). It makes a huge difference and stops your hard work looking like a high school project.

When I examine a viva, I look for four key components:

(i) The basics. Has the candidate realised the basic science/concepts underpinning their thesis and research and can they discuss them and answer questions on them? What about theory: can you explain key theoretical results that are relevant to your topic?

(ii) Experimental. If the work is experimental, do they understand the techniques, the theory behind them, and the limitations of their use/application? What about the equipment they've used? It's mostly electronic/digital these days, but are there issues around signal filtration, data acquisition, digital vs analogue recording, etc. that might have an effect?

(iii) Computation/simulation. Not so much an issue for the 'serious' theoretical projects, but it's quite usual for us experimentalists to use a computer simulation to shed light on our experimental results. But what about that simulation program? How does it work? Has it been properly validated against analytical results? Validation is really a key point.

(iv) Understanding of the limitations (nothing can be perfect). I think most candidates view the viva as a sort of 'exit interview', in that it completes their PhD and earns them the right to the title of Dr., and to graduate etc. That's obviously true, but it's also an 'entry interview'. Once 'doctored', the candidate is essentially qualified to be a professional researcher: maybe a post-doc in charge of PhD students of their own, or hold a fellowship, academic post etc., so whilst the viva is about checking the quality of the PhD work, it's also about checking the ability of the candidate to go forward into their future as researchers and academics. I think key to doing that is to reach the point in the viva where the candidate can discuss (safely) what they feel the limitations and strengths of their work is. Essentially, it's the 'If I had to do it all again, how could

I do it better?' question. Once they leave the viva having passed, the only thing separating the candidate from their examiners is age and experience, we can never produce a perfect piece of work; we just try to learn from our mistakes and experience to make the next piece of work a bit better.

It is important to note that the letters in front or after one's name do not make you a fantastic scientist or medic or whatever, so being able to spot strengths and weaknesses in your own work and practice, is, I think, a key part of reflection or self-insight that is important in maintaining professional integrity of your research.

I think you cannot prepare for a viva. It's not an exam, so you do not know what the questions are going to be. I have real reservations about the practice of having a 'mock viva' (a new directive from my own Institution), as I think it drives a certain mentality that there are a limited range of questions to prepare answers for. By all means discuss the strengths/weaknesses of your work with your supervisor, and the sorts of topics that might come up, but never get too fixated by the exact question. In one sense there is no defined syllabus in the viva, but in another sense there is: it's your thesis and all the science/learning that underpins it.

Here is some advice as you prepare for the viva:

(i) You should certainly read your thesis carefully, for this I would suggest not looking at it for a couple of weeks after you submitted it and try to read it 'afresh'. Make sure you read what's actually written on the paper and not what you think you wrote—we all know what we're trying to say, but not always do we write as clearly as we think. Remember your written thesis is all the examiners have to judge you on. Also, remember that the primary purpose of the viva is to satisfy the examiners that you performed the research and wrote the thesis.

(ii) It is important when you are arranging the viva, to be careful how you communicate with the examiners. The examiners will find a suitable date and then contact you directly/indirectly to ask you to confirm your availability. If you have any obvious dates that you know you cannot be available for, then you should tell your supervisor to communicate these to the examiners as early in the process as possible. Depending on local arrangements/customs, the proposed date/time of your viva will be communicated to you, either: (i) directly from the internal/chairperson; or (ii) via your supervisor. Sometimes the external might, but in my experience that's rare. It's best to respond via the same channel that you were invited (copying in the same addressees). This sounds like stating the obvious, but where you need to be careful is that generally you should NOT be contacting examiners directly unless you are responding to them—it could be interpreted as attempting to influence them, which is a no-no. Some examiners are very sensitive to this, others don't care, and the formal regulations about contacting examiners also vary from institution to institution. If in doubt, check with your supervisor.

(iii) It is vitally important that as a candidate you find and read the exam regulations for your institution regarding PhD degrees. It will show you the requirements for passing, and the variations of minor/major corrections and revise/resubmit. They are generally similar in the UK, normally requiring 'evidence of originality and contributing new knowledge to the field of study', but there are some differences. This might include the standard of the work (e.g. does it have to be of publishable quality, or merely capable of being published in some form at some point?), or evidence of appropriate industry/effort (i.e. does this look like it took three years, or could someone have done it in 12 months?) and similar wordings. There might be specific details about the examination too. Some institutions have a chairperson present as well as internal/external examiners, and for some types of doctorate you might have two external examiners as well as an internal, etc. It's always good to know what to expect. Some departments might require you to give a short presentation or lecture on your thesis immediately before the viva, with the examiners and staff/postdocs and PhD students able to come as the audience. In the UK, such lectures are 'no questions' affairs in my experience, but obviously overseas there can be variations, such as the viva following in the same room with the audience remaining to observe.

You should read the regulations long before the viva to be honest, but it's never too late and you should be asking "what to expect" questions to your Supervisor and post-docs too.

(iv) Make sure you can summarise your thesis in 5–10 min, as this will probably be the first question you're asked in the viva. We ask that one to help the candidate settle their nerves. For this, be able to 'tell the story'… what is the thesis about, why is it important, what are the key findings, how do they fit together, etc. Don't go trundling out 'In Chap. 1…. In Chap. 2….etc.'

(v) Be confident in your work. Remember you are the only person in the world who has done your actual experiments, analysed your data, and written your thesis so you are the world expert on your PhD. But don't be cocky or arrogant; that will get picked up in the opening minutes and generally elicits the response of 'right, we're going to take this guy down a few pegs for the next 3-4 h' from your examiners.

(vi) Make yourself comfortable. Some universities have dress codes, others not, but you want to dress smart enough to look professional but still be comfortable. Have breakfast or lunch beforehand and bring a bottle of water in case the examiners forget to get one ready for you. Most vivas will last 2–4 h, so you're going to be there for some time. Don't forget you can ask for a short break, and usually the examiners will offer one if it goes past 2–2.5 h. If a break is called (usually 5–10 mins), the viva stops—so if one examiner needs to pop to the toilet, the other one can't continue the questions/discussion alone. They might switch into small talk though, so don't be surprised if the guy who's just been doing a fair impersonation of Torquemada for 30 mins grilling you on Chap. 3, suddenly smiles and asks you if you did anything nice last weekend.

(vii) The length of the viva depends on several things. It's not as simple as the better the quality of the work, the shorter the viva—although I did once run out of questions after 1.5 h because the thesis (and candidate) were so good. I've been in plenty of vivas that I expected to last 2–2.5 h but have run to 3–3.5 h simply because the three of us got carried away talking about the thesis and lost complete track of time. Your examiners are human, though, and eventually get hungry too—which can place a backstop to the viva. If the staff canteen stops serving lunch at 1.30 p.m., then the viva will have to finish by 1 p.m. at the latest, then 15mins for the examiners to congratulate you and send you off to celebrate, then fill out the paperwork together, leaving 15 min to hotfoot it to the canteen before it shuts. Similarly, some examiners in the afternoon sessions will aim to catch (i) train home, or (ii) closest bar by 5.30 p.m., depending on their priorities (!). Of course, if there's good reason to miss lunch/opening time, your examiners will keep going… so don't rely too heavily on the clock.

(viii) Try to enjoy the viva. The first twenty minutes will seem like an age, then once you get into discussing your work, you'll probably find the next 2–3 h will fly by. Whilst there's obviously an examination aspect to it, it should rapidly turn into a discussion on the work, everyone relaxes and (hopefully) enjoys it. Don't forget this is probably the only time in your life you'll get to discuss your research with anyone for any length of time without them: (i) falling asleep; (ii) checking their smartphone; or (iii) starting to watch the TV. Unless you talk a lot to your cat or dog, and that doesn't count, because I am sorry to burst your bubble, but they, like your partner/family member, are only to pretending to pay attention after the first three minutes because they love you."

Professor Katherine Berry is a *Clinical Psychologist* and a Professor of Clinical Psychology within the Division of Clinical Psychology at the University of Manchester and here Katherine shares her experiences of examining vivas:

"It is hard to say exactly how many vivas I have examined as I have worked as an internal and external examiner on ClinPsyD programmes which easily increases the number of vivas you do a year. In terms of PhDs. I would say an average of three per year since 2013 when I first became a senior lecturer.

I have probably equal experience of internal and external examining. With internal examining, I would naturally take more of a back seat and let the external lead. However, I have been an external examiner a couple of times, where the internal examiner dominated the session (one in a positive way as the thesis was really good and one in a more negative way as the thesis was weaker and the internal examiner had a real issue with it).

I have throughout my career always been in my area of expertise for PhDs but sometimes ClinPsyD vivas are not within my immediate area of expertise particularly when I am the internal examiner.

The most enjoyable aspect to being an examiner is reading the thesis and learning new information. It is a rare opportunity to read about a topic area in depth and it is good in terms of keeping me up to date with current papers in the field.

Minor corrections are definitely the most common outcome within the theses and vivas that I have examined. Some examiners can hesitate in giving major corrections because of the psychological impact of this on the student, but sometimes it is the fairer option in terms of allowing them adequate time to make the changes that they need to make.

In a thesis I am looking for the addition of new knowledge to the field, rigorous methods or, in the absence of absolute rigour, an acknowledgement of limitations.

In a viva I look for ownership of the work and decision-making. An ability to talk around the topic and justify decision-making. Thoughts in relation to practice or future research implications that arise directly from the thesis and not more general implications relating to the research topic.

To PhD candidates, I would just suggest knowing your work well, including the rationale for the decisions you made. I think you can spend ages trying to second-guess what questions will come up and you could be wrong. I could question to prepare would be a summary of the thesis or an explanation of how you came to study the topic as, in my experience, exams always start this way.

It is a very emotional experience for students, and I think it is lovely to see people's relief at passing with minor or no corrections. Conversely, there have been some tricky experiences when the examination process hasn't gone as well as the candidate might have liked. I think it is also interesting when there is a mismatch between the quality of the thesis and the performance in the viva which can go both ways, i.e. excellent thesis and weak viva or vice versa."

An experienced examiner at a UK university shares his experiences of examining PhDs:

"I have examined over 30 theses and have been both the internal and the external examiner. The experience differs depending on whether you are an internal or an external examiner. The internal examiner is responsible for liaising with all parties, organising the viva and overseeing completion and submission of viva-related paperwork and reports. The internal examiner also asks questions during the viva. The external examiner leads the viva discussions from the academic point of view, with input from the internal examiner. Both examiners assess the quality of the thesis and candidate and decide on the outcome. I have over the years, examined PhDs in my specific area and on broader topics. When I examine PhDs I enjoy learning about the research, how the candidate approached the research and how the candidate addressed problems that arose, as trouble shooting and overcoming hurdles is a key aspect of research.

In my experience minor corrections is the most common outcome. In a thesis I am looking for a well-written, coherent body of work that is well structured and well presented.

In a viva I am looking for an ability to discuss the findings pragmatically. Also, I look for an understanding of how the candidate's work relates to the research in the field. An ability to answer questions on all aspects of the thesis. An understanding of the limitations of the work.

My top tips for PhD students preparing for the viva are to know the thesis inside out. Also, to know what research has been carried out in the area and know the methods/results/conclusions of key papers published in the literature. Put enough time aside to prepare properly.

The most positive experiences are gained when candidates are extremely well acquainted with their research and how the research fits into their particular field. Candidates should be able to comment on what they would have done differently. Candidates should listen carefully to the question that is asked and answer that specific question, not another, albeit related question, that they know the answer to."

Summary

- Every viva experience is different; even within the same subject area and same institution they can vary wildly.
- The viva can leave you with both positive and negative memories; either way, it is a unique experience that you can learn from.
- Examiners are there because they are interested in what you conducted and want to discuss your exciting research with you.

5

Making the Most of and Enjoying your Viva

Stacey Bedwell

5.1 Making the Most of and Enjoying the Viva

In this chapter we intend to highlight what you can do to make the most of the unique opportunity presented to you in the doctoral viva, and what you can do to ensure it is an enjoyable experience. It is inevitable that you will feel a little nervous about your viva exam, that's perfectly normal and, quite frankly, it would be concerning if a candidate was not even slightly nervous about it. Additionally, there are going to be some aspects of the viva process you don't enjoy. Some of us will enjoy it more than others; that's also fine, we are all individuals after all. In this chapter I will try to help you avoid with the content in this chapter is missing the opportunities the viva brings and missing the chance to enjoy something you could have.

5.2 An Enjoyable Experience

The doctoral viva itself can be enjoyable to an extent. I would, however, be mindful of any accounts from academics who say that they enjoyed absolutely every second. I personally don't think that's realistic and is likely an unobtainable goal to set yourself. However, there is no reason why a large proportion of your viva exam cannot be an enjoyable experience, and there is certainly no

S. Bedwell, I. Butcher, *How to Excel in Your Doctoral Viva*,
https://doi.org/10.1007/978-3-031-10172-4_5

reason why you should not be able to enjoy both the build-up to the event and the aftermath.

The Build-up to the Viva

Most situations in life are less intimidating if you remove the unknown. There are some unknown parts of the viva over which you will have no control. For example, no matter how much preparation and practice you do, you can never know exactly what the examiners will ask. There are some parts you do have control over, though, like knowing about your examiners. In the age of Google and social media it has never been easier to do your homework on your examiners before you arrive for your viva exam.

Reflect on What you Have Achieved

In the run up to the viva it can be easy to develop something of a tunnel vision attitude, forgetting about all the fantastic things you have achieved in the previous few years of your PhD. It's valuable in the weeks and days ahead of your viva date to be able to take some time to think and to reflect on what you have achieved. At this point you have written a complete PhD thesis; this is a great achievement and something to be proud of, regardless of the viva outcome. You likely have many other achievements to reflect on too, both small and large. These might include international and national conference presentations, journal publications and all the new skills you have developed as a researcher. You might have also made new friends and had many new experiences along the way.

Reflect on why you Took on this Project

Most of us take on PhD projects for a reason. It does not tend to be a journey that is embarked on without thought and consideration. In some cases, you proposed the project yourself, perhaps based on a personal interest or experience. In others, you applied to a project because it caught your eye, perhaps as something new and exciting or as something you saw as valuable. Whatever your reason for starting the PhD all that time ago, you might have forgotten or lost sight of it on the way. Take some time ahead of your viva to remind yourself of what initially made you so excited about your project. Of course, part of the reason might have been to get that Doctor title, but it won't have been the only reason. Regardless of how consuming and important your viva might feel at this point, passing the viva is not the sole goal of your PhD and the research you have done. Reminding yourself of this can help you take the emphasis of the exam process and enjoy the viva for what it is really all about. It is a chance to discuss your research with other experts.

Look at a Wider Perspective

Part of the reason the viva and the build-up to it can feel overwhelming and unenjoyable is because we can become so focussed on it. If all we can see is the viva and our focus on passing the viva, we lose sight of the bigger picture. This brings us back to the earlier point of reminding yourself why you took on the project in the first place. Try to step back, look up, and look at your PhD and viva in a wider context. The bigger picture is that you have contributed to the state of knowledge and understanding in your field. That remains true no matter what happens in the viva. If you have publications, if you have presented at conferences, if you have made connections with other academics in your field, all of these achievements and contributions remain true and relevant regardless of the outcome of the viva. It is also important to remind yourself that the viva exam isn't the only chance to prove yourself, even if this is sometimes how it is sold to us. The truth is that some candidates do fail the viva; I won't pretend it doesn't happen, but I will also tell you it is never the end of the world and that it has little to no impact on most people's following academic careers. There are always second chances in most endeavours.

Prepare for Tricky Questions

One of the most useful things you can do to help yourself into an enjoyable viva is to not bury your head in the sand. If there is any aspect of your work that you dread coming up, the worst thing you can do is to simply hope your examiners don't ask about it. Of course, you might be lucky, and it might not come up, but you will inevitably spend your entire viva thinking it might and anticipating not being able to answer. It's impossible to fully enjoy the discussion about your favourite part of your thesis, that aspect that usually gets you excited, if in the back of your mind you are wondering when and if they will ask about that thing you really don't want to talk about. Instead of living in fear, in the weeks before youe viva you should set aside some time to really get to grips with that tricky subject. You don't necessarily need to learn complex theories and concepts which have little relevance to your work; if they do come up, however, you want to have an answer that's better than 'I don't know'. It's important you can demonstrate a willingness to consider things outside of the constraints of your thesis and, similarly, a willingness to admit when a topic is beyond your expertise. If there is a topic you are specifically worried about, it's a good idea to sit down with your supervisory team and talk through how to approach it. Having a clear strategy will make it much less intimidating and will also allow you to relax and enjoy the parts of your research you are passionate about.

Own your Work and Ideas

One of the main purposes of the viva is for the examiners to confirm you did the work yourself and also that you wrote the thesis. It's vital that you can demonstrate that it is your work, not only in the sense that you wrote it but also that it is a product of your decisions and thought processes, rather than, for example, those of your supervisor. To demonstrate your ownership of the research, be prepared to justify all of the decisions and conclusions made and explain the thought processes behind aspects such as experimental design. You should also be prepared to defend those ideas and conclusions.

Practice Receiving Constructive Criticism

Although it is vital to own your work and to be able to defend it, it is equally important to be able to take criticism on board. This is all part of being a researcher; part of your viva is demonstrating to your examiners that you have developed into an independent researcher. Taking criticism does not mean to blindly accept the opinions of your examiners, however. Remember that they might be experts in the field, but that you are the expert when it comes to your research project. Be prepared to listen to, digest and discuss critical points.

If you are a person who struggles with criticism or who tends to get defensive, you might find it helpful to seek critical comments on your research before your viva. This could be in the form of a mock viva, feedback from your supervisors or even going through the peer review process with a journal submission.

Practice Explaining your Research

If I were to tell students to do one thing before their viva it would be to practice explaining their research to varied audiences.

Prepare for the Logistics of the Day

How important this point is depends, to a large extent, on the type of person you are. However, the most easy-going of us can become a little flustered on a stressful day. So, to avoid any logistical errors or extra things to worry about, prepare for all the practical elements of the day in advance. This includes ensuring you know details like the time and location of the viva well in advance. If your viva is planned to take place in a room or building you are not familiar with, it may be useful to make a visit to the location before the viva day itself. This way you remove any chances of getting lost on the way to your viva and you can familiarise yourself with the room layout beforehand (if possible). For university campuses with high security, you can also that check your ID card gives you access if it's not your usual place of work. Other venue things to check out beforehand could include things such as the location of the nearest toilets to your booked room and also if you need any adjustments for specific needs.

Depending on the time of day your viva is being held, you may want to give careful consideration to when you will arrive on campus and how you will travel there. If you are the kind of person who worries about being late, sitting in unexpected rush hour traffic on your way to campus could be something you want to avoid.

The Viva itself

You may find that once you get into it and relax a little, your viva is much more enjoyable than you might have anticipated. There are a few things you can do whilst in your viva to ensure it remains a positive experience.

Take Breaks

Every person I have spoken to about this did not take a break during their viva, whether it was in person or online. In hindsight, I did fine without a break, but I do think I should have taken one. There is no reason why we need to do the whole viva in one go. Talking constantly for three or four hours is quite exhausting and I do think my response to some of the questions became a little more defensive towards the third hour mark in my own viva. I think it's especially important to take a break and regroup your thoughts if you have come to any kind of disagreement with your examiners. As I have said elsewhere, disagreeing is fine, but you don't want that feeling of differing opinions to carry through to the rest of the discussion. Sometimes it is beneficial to take a short break, get some fresh air and come back to the table refreshed.

Considering online viva exams, in particular, 'Zoom' fatigue is a real issue (Shoshan & Wehert, 2021). The way you engage with a conversation on screen is quite different to real life and can result in increased tiredness and headaches. Avoid this effect as much as possible by taking frequent breaks.

Don't Make it Harder for Yourself

The viva is not a test of memory. It's already quite a cognitively demanding task, so there is no need to make the process more difficult for yourself by making it a memory test too. Rules vary between institutions, but every viva set-up will allow you to bring in a copy of your thesis with notes. If you know there is something you are going to struggle to recall or want to remind yourself of, write it down. There is no need to memorise trivia.

Ask for Clarification

Throughout the viva discussion itself, it's helpful to try and remind yourself that it is simply that, a discussion. It is not a quiz or a set exam. As the viva is intended to be a conversation, it is perfectly fine if you need to, or want to, ask your examiners to clarify what they mean by a certain question or point they make. Similarly, they might ask you to clarify something too! The whole process is meant to be focussed on gaining a clear understanding of the details within the project you have produced. Seeking clarification is useful for everyone involved. There is no need to feel like you must answer the question immediately; your examiners would usually rather you are able to confidently discuss a question where you fully understand what they are asking.

Drink Water

The viva can be long. Often lasting around three hours, this can be a very long time to go without water. Unfortunately, many candidates do forget, or simply fail, to drink water during the in-depth discussion. This is either because they become so engrossed in the conversation, they don't want to pause, or because they feel they don't want to have to take a bathroom break. Dehydration can cause headache, fatigue, irritability and even impact on cognitive performance, so it really is in your best interests to have a brief sip of water from time to time. Furthermore, it's a well-known trick to stop and take a sip of water if you want to take some time to think about what you are going to say—or to encourage yourself to stop and think before you speak.

Take your Phone

This final tip to consider during the viva itself may sound like quite an odd one. I'm not recommending you take your phone out and use it during the conversation, but it will come in useful when you come out and you are waiting to be called back in to hear your outcome. Those few minutes could feel like a long time, and you might find you want to tell someone you have finished and are waiting. This could be family, a friend, a colleague, or your supervisor. It is worth noting that, depending on your institution, you might not be allowed to take a phone in with you. You should check this beforehand.

The Aftermath

When people ask about my viva and how I felt afterwards I often say it was all a very anticlimactic experience. Leaving the viva exam, whether in person on online, rarely gives the sense of closure you might be expecting, and which you will likely be used to from final exams in undergraduate and postgraduate taught study. It's a very odd feeling, because you have finished and passed, but in most cases not fully—when you walk out of that room most of us have some corrections to do, so we don't technically have the PhD yet. Despite this technicality, I think it is important to mark the occasion and celebrate. Perhaps most importantly, I think every candidate should celebrate, regardless of the outcome. Even if you must resubmit or resit the viva, you have still completed a huge undertaking and done the viva exam. Many PhD students build up the viva in their minds beforehand, such that it becomes this all-consuming thing hanging over them. Because of this, if you don't properly mark the occasion, it is not going to feel like you ever got to the end.

Regardless of outcome, completing and defending a PhD project is a very impressive achievement. By this point, you will have put a huge amount of work into this over many years. It is worth celebrating in some way, whether that is with your supervisory team, colleagues, family, friends or even if you simply do something to mark the occasion yourself. You don't have to throw a huge party—just do something special to mark the achievement.

5.3   The Value of the Viva

It's entirely understandable that the value of the viva exam, beyond it being a means to an end regarding your doctorate, is hard to visualise. There really are, however, a lot of reasons why the viva experience is valuable—not only for your academic career but also in other areas of life.

A Chance to Show off

The PhD or doctoral degree experience can be a lonely one. Granted, some PhDs work in large research groups and get to share their exciting work and findings on a regular basis; for instance, at weekly lab meetings. However, this is not the case for all candidates. My own experience, for example, was quite a solitary one. I did spend a much more than average amount of time with my supervisor (with hindsight I do realise now just how fortunate I was in this respect), but very few other people. I did not work in a large research group— it was literally just me and my director of studies. For PhDs like me, who didn't get the opportunity to share findings and discuss research details much outside of their direct supervisory team, the viva is a really valuable opportunity to show off all that great work you've been doing. For those of you who did present regularly at lab meetings, however, the viva is not wasted on you; this is your time to shine outside of the group that have closely watched your work grow over the past few years and gain a new perspective.

A Chance to Discuss Critically

We often talk about the importance of criticism in in science, yet academics rarely have the chance to focus on the critique of one specific project in such an intense way as the doctoral viva. There is value here not only for you, but also for your examiners. Most doctoral researchers are very passionate about their work and the contents of their thesis; they should be having put so much effort into it. It can be disappointing, therefore, to be presented with a lack of the same level of enthusiasm when explaining findings and implications to lay people. For example, in the run-up to your viva, as useful as it is to practice explaining your research to anyone who will listen, your non-academic friend, or even your colleague working on a different project, is unlikely to find your work as thrilling as you do. This is completely understandable. Your examiners, however, even if they don't work on the same topic as you, are going to have more of a vested interest in what you have done and what it means, perhaps even in relation to their own developing research. This common passion is valuable in your viva. It enables prolonged and in-depth critical discussions that you would not get with anyone else, except perhaps your supervisors.

Gaining Valuable New Perspectives

I've mentioned previously how the PhD experience can be a solitary one. It is very easy to fall into a PhD bubble, where you communicate only with your closest work colleagues. The viva can also be viewed as a networking opportunity, in the same way as presenting your work at an academic conference. Up until now, most of the outside input to your work has come from your supervisory team. For some, the supervisory team really means just the director of studies if the other supervisors offer expertise outside of the specific scientific methods and approaches. This can result in the development of quite a narrow view. Of course, we all read widely and constantly consider the literature in our research development, but our supervisory team, essentially our research group, naturally becomes the strongest influence and offers just one or two perspectives.

The viva is a chance to bring in new perspectives. Usually, your examiners have not been involved in your project at all right up to the point of reading your thesis. This really does offer a fresh look on your approaches, methods and interpretations. A very useful aspect of the way the examiners are chosen is that we usually choose experts, but institutional rules stipulate they cannot have been involved in your progression or the project before. This might seem like a difficult task and may even bring you to question why; surely the best experts in your very specific field have already been involved in some way. However, this common rule is actually there for your advantage. Reading a completed project without knowledge of how it may have changed, what was discussed in lab meetings or what happened at the MPhil. Transfer, enables the examiners to bring their own, often different, perspective to your work. Often this can open up new ways of thinking about and interpreting findings and conclusions, expanding and building your research moving forward.

Practice in Articulating Complex Ideas

Although the viva is considered the final hurdle to your PhD, it is just the beginning when it comes to your career. Of course, its main purpose is to defend your doctoral thesis, but it has other uses, especially if you are pursuing an academic career. Many PhD graduates go onto roles as lecturers in universities. This can mean a big leap from working in a very small tight-knit team of experts to explaining complex ideas and theories to very large groups of undergraduate students and non-experts. PhDs are often very used to conversing with and discussing scientific concepts with other researchers. It is easy to get used to people knowing what you are talking about and not needing to break things down or simplify. Explaining complex ideas in accessible formats can be challenging and is an acquired skill. For many recent PhD graduates who go onto lecturer or teaching fellow positions, they are very much thrown in the deep end. Postgraduate teaching qualifications are usually undertaken alongside the first year or two of a full-time teaching role. Your viva and the preparation you do building up to your viva can help make that transition easier; all the practice explaining your research to lay audiences and articulating complex ideas to your examiners will set you up well for that first lecture.

Your viva skills in explaining complex content is useful outside of academia too. Whatever direction your postdoc career takes, there are very few jobs in the world that don't require good communication skills. Your viva success is evidence of yours.

Practice in Selling you, your Work, and your Skills

Modesty seems to be a common trait in academia. When it comes to being able to sell yourself and your abilities in job applications, interviews, and grant applications, being able to step past the modesty and really sell your unique skillset is valuable. The viva experience helps you to achieve this. As part of the viva you will be essentially selling your work and its value to your examiners. You will likely, at some point, have to explain the real-world implications of your research, the value of your findings to society or the impact you have made during your PhD. Here you are selling not only your research, but also you and the skills and knowledge you have developed to produce it. Articulating this in a passionate and convincing argument, through your viva, is great practice for your future career.

5.4 The Purpose of your Viva

Believe it or not, the viva is not designed to make doctoral candidates feel uncomfortable. It's also not intended as an opportunity to put you on the spot and hope you crumble under the pressure. It's not a lie that academia is generally a competitive world. Getting onto your programme was competitive, getting funding is competitive, awards, grants, promotion, jobs, publication, it's all full of fierce competition. So naturally, we are used to being in competitive environments; indeed, we might even thrive on competition. The viva, however, is one of the few academic milestones that is not competitive in the way we are all so familiar with. It's no surprise that we might default to feeling competitive about the viva, because it probably comes as second nature when you get to this point in your academic career.

If the purpose of the viva is not competition, what is it then? Well, it's your time to shine. There are several main purposes of the viva, and reasons why the viva is an oral exam, structured the way it is.

- To demonstrate you wrote the thesis.
- Without the viva, technically someone else could have written your thesis; your examiners would never really know. To be able to discuss your thesis and its contents in the great detail covered in a viva, it must have been written by you.
- To demonstrate you did the work.
- To be able to write about research is one important skill, but to achieve the PhD or doctorate you also need to have carried out the research yourself. As with the writing aspect, to be able to discuss your research, justify all the decisions and explain its complexities in the level of detail of a viva, you have to have completed the research yourself and understand what you did.
- To defend your work.
- There's only so much defence you can portray in a written format. Without the back-and-forth conversation nature of a viva exam, it's simply not possible to defend your work in the same manner. The viva gives you the opportunity to do this.

5.5   How the Viva Helps you Post-PhD

In this chapter we have established that it is important to try to avoid developing tunnel vision when it comes to the PhD viva. One prominent purpose of the viva exam is a means to an end; there is no denying this. But it is valuable to acknowledge how the viva and the whole viva process, including your preparation, can help you in your career post PhD. This applies whether or not you intend to stay in academia. There's a wide range of areas in which the PhD viva can help you:

Job Interviews

Whether you wish to remain in academia or pursue an entirely different career post PhD, you are going to have to go through an interview process. Job interviews can be daunting, regardless of the field. However, having just gone through a PhD viva, you have essentially experienced what is essentially a very in-depth interview. There are more similarities between the viva and a job interview than how it physically looks. A large part of the viva involves you proving both yourself and your work's merit to your examiners. This is the very nature of a job interview, so at this point in time you are really quite practiced.

If you are interviewing for academic roles, especially those involving research, you are inevitably going to be asked to explain your research to the panel. This is exactly what you have recently perfected in preparation for your viva, so there is probably no time you will be more confident in your ability to do this.

If you are interviewing for non-academic roles, this doesn't mean your viva preparation efforts are irrelevant. There are very few jobs that don't require an ability to communicate with varied audiences effectively, and your viva performance demonstrates your ability to do this well.

The Workplace

Your viva preparation and experience does not stop being useful once you are appointed and find yourself in the workplace. The need to communicate with various audiences will continue to be important in team meetings, stakeholder meetings, presentations, consumer-facing roles and a range of other circumstances. Your skill in explaining complex concepts and ideas could aid you in developing new ideas in your role and in your own progression. Workplace relationships are also vital in any career, part of which is based on clear communication of ideas between colleagues.

Teaching

Many PhD graduates go into academic teaching roles, such as lectureships. Often this happens immediately after completing the PhD, sometimes even beginning before the viva has taken place. One of the realities of academia is that new lecturers are often thrown in the deep end. Although most universities require those new to the role to complete a teaching qualification, such as a postgraduate certificate in higher education, this is usually completed alongside the first year or two of teaching. This means that when a newly appointed lecturer delivers their first class, they are often doing so blind, with little to no training at that point.

Teaching effectively, at any level, requires an ability to explain complex ideas and theories to non-experts in an accessible and digestible format. This is where your viva experience comes in handy. You have spent a considerable amount of time and effort fine-tuning your skills in doing exactly this. By the time you come to your viva exam, you can articulate your complex research to an audience of non-experts. This sets you up well for those first few months in your new teaching post.

Conferences

An important element of an academic career is communication of ideas and research to wider audiences. This includes conference presentations. The viva is a fantastic experience and a sound training for giving engaging and inspiring conference talks and poster presentations. In the build-up to the viva, and also in the viva exam itself, you have been refining your skills in summarising your work, explaining its impact, and essentially selling the value of what you have done to your examiners. At a conference your audience might be much larger and more varied, but the premise is the same. Your goal in a conference presentation is usually to summarise years of work in around ten minutes whilst engaging your audience and convincing them of its value. You are quite practiced at this by the time you walk out of your viva.

Another key element of conference presentation is the question portion. This is a time for other academics, who may or may not be experts in your field, to ask questions about your work. As the presenter, you can pre-empt what audience members might ask, but you cannot guarantee what will come up. This should sound very familiar, because it's very similar to your viva. Many academics find that after completing their PhD the question portion of a conference presentation becomes much less intimidating. This is likely a little to do with increased confidence and experience at presenting, but also a product of having gone through the viva process.

Meetings and Presentations

There are a number of ways the PhD viva process prepares academics to shine in future meetings and presentations. Nothing quite prepares you for criticism like the doctoral viva process. As academics, we are under constant scrutiny. This certainly does not end when the PhD is complete, the process of going through the viva could actually be considered as practice, for being able to take criticism on board, not taking it personally and using critique to improve future work and research. The viva is a process of defence in itself, so it naturally builds your confidence in defending your ideas. This can be a valuable asset when it comes to decision-making in team meetings or defending your ideas when presenting them, either in small groups or at big international conferences.

The viva experience is a small step in starting to build your academic network. This might be one of the first times you have discussed your research in detail with experts outside of your direct research group or lab. Beyond your viva you can begin to expand this network of academic relationships, to continue to build your research career and open up new opportunities.

Finally, the viva gives you a great opportunity to practice at performing under pressure. This is often an aspect of the viva that makes people feel nervous—the requirement to think on the spot and answer a question they might not have been prepared for. Despite this, it is great practice at a valuable skill you will likely need at some point in your future career, whether that's in academia or elsewhere.

5.6   Favourite Parts of the Viva

Hindsight is a wonderful thing, and often allows us to see positive aspects of a situation after the fact, things that perhaps we did not notice as enjoyable at the time.

I would be lying if I didn't admit that my favourite part of the viva was when it was over. People often describe a feeling of relief after being awarded their PhD, and this is exactly what I felt. Sadly, the whole experience was a bit of an anti-climax. I think this is largely because a PhD and the associated viva exam is a very lonely experience. You will never get the same celebratory experience as you may have at the end of final exams in previous degrees, because you are never going to pass your viva at the same time as others in your cohort. Despite the anti-climactic feeling, the end, when the examiners told me I was being awarded a PhD, is the part I recall most vividly and the time when I can say I felt confident in the work I had done. Before that moment, there was always some self-doubt. When I say that the end of the viva was my favourite part, it's not necessarily just because it was over. It was a little disappointing to think that was it; the project was done, what had become my identity for nearly four years had come to an end. I think this was my favourite point in my viva experience because it was the time when I felt like I had become the expert.

5.7 Positive Viva Memories

With so many emotions often running high after a viva, and after what has likely been quite an intense few hours, it's all too easy to focus on the negatives. Naturally, there will be some parts of the viva experience that you have no interest in reliving, and also perhaps questions you feel you could have approached much better in the moment. It's also very easy to talk about the negative or less enjoyable aspects of the viva experience when others ask about it. It's very much our human nature that we highlight the negative and don't bother so much with communicating the positive (just look at Yelp and Tripadvisor reviews!). This can give current PhD candidates a negative view of something that really was much more positive. When asking recent, and not so recent, PhD graduates about their experience, it might help to specifically ask them if they can recall a positive memory of their viva. You might be pleasantly surprised…

Giulia Furini, who completed her PhD in Biosciences, says *"More than pin-pointing a specific memory, I mostly remember the positive atmosphere of the viva itself, the challenging and enjoyable discussion, and the feeling of being quite confident about what I was talking about. I think I was very lucky with my examiners, as they made me feel good about my work and treated me like a peer, despite the obvious difference in knowledge and experience between me and them."*

Damien Neadle completed his PhD in Animal Communication. He recalls *"Looking back, the experience was a generally positive one, but the moment that I first defended my definition of culture and both examiners accepted it and retracted their concerns was the moment I realised I could do it."*

Theodore Hughes-Riley completed a multidisciplinary PhD in 2014. He says that *"Looking back I think that the viva itself was actually okay, but I honestly don't remember how I really felt in the viva itself, other than nervous."*

Nicole Pearcy completed her PhD in Mathematics. She says that *"I remember my internal examiner coming to collect me before the viva—he immediately said he really enjoyed reading my thesis and was looking forward to the discussion, or something similar, which was really nice to hear and boosted my confidence going into my viva questions."*

Richard Fereday completed his PhD in Psychology. He recalls *"The overall positive atmosphere. There was no stress, the questions were asked in a friendly manner, and it was not a 'grilling'—it was simply a chat about my knowledge. I do remember saying to the chairperson that a beer might be nice. He replied, saying there might be can of lager somewhere. I declined—I prefer IPAs."*

Summary

1. The viva can be an enjoyable experience.
2. The value of the viva is far more than just achieving your PhD.
3. You might be surprised by the positive memories you come away from your viva with.

Reference

Shoshan, H. N., & Wehrt, W. (2021). Understanding "Zoom fatigue": A mixed-method approach. *Applied Psychology*, *1–26*. https://doi.org/10.1111/apps.12360

6

After the Viva

Isabelle Butcher

Often so much time and credence is given to the viva itself and preparation for the day, that what happens following the viva is not considered by candidates. This leads to candidates not being fully informed of the process following the viva.

It is critical to note that there is not a numeric pass mark—there is no marking rubric or marking criteria which you as the student are examined against. Your examiners do not add up a score for your performance in the viva. There are, however, a set of outcomes all which do depend to a degree on the HEI. It is important that prior to your viva examination you ask your graduate school for a copy of the possible outcomes. Each HEI has these available to ensure the process is transparent.

As mentioned in previous chapters, unlike many other similar scenarios, you do find the outcome of your oral examination and PhD on the day itself and from the examiners. This is a scenario that many of us are often not used to and so can lead to candidates experiencing dread or fear because they feel as if everyone is resting on their performance on one day.

Your examiners will ask you to leave the room after your viva whilst they discuss their decision. This can often happen in just a few moments, but it may take longer. It must be remembered again that the length of time you are not in the room is no reflection of your performance or your outcome. It is advisable to stay nearby and to wait calmly—the temptation, of course, is to

© The Author(s), under exclusive license to Springer Nature Switzerland AG 2022
S. Bedwell, I. Butcher, *How to Excel in Your Doctoral Viva*,
https://doi.org/10.1007/978-3-031-10172-4_6

ruminate on your performance in the examination and to communicate with peers on phone; however, this can only serve to create greater stress. This is often why vivas in person happen in a neutral room so you are not waiting in a corridor by your supervisor's office and near to peers. When they are ready the examiners will call you back in, at which point they will let you know if they are recommending you are awarded the degree of a PhD, and whether or not this is dependent on any corrections or revisions to your thesis.

6.1   Outcomes

Whilst reading this section please check with your HEI for the exact guidance and breakdown—this often varies from academic year to academic year and from institution to institution. The terms 'corrections' and 'revisions' are used interchangeably as some HEIs prefer to refer to 'corrections' whilst others opt for 'revisions'. These words often refer to changes that need to be made to the thesis and the doctoral work to ensure it is improved. Table 6.1 shows a summary of possible outcomes. You will be advised by your examiners which award applies to you.

6.2   Corrections/Revisions

If you are asked to complete corrections, these are sent to you, most often by your graduate college or via your internal examiner, in a list which the external and internal examiner have created and agreed upon. You may receive your corrections soon after your viva, or it may take several days. It is important that you do not begin making changes to your thesis until you receive the corrections in an email or letter, even if some points were mentioned in the viva. The corrections may include items discussed in the viva, but there may also be other minor corrections. These can include, but are not limited to the following: typographical errors, citing references that have not been included, making research decisions more transparent, writing new methodology sections, ensuring that the thesis story is coherent, and ensuring that each aspect of the thesis fits together.

Students are often concerned about the difference between minor and major corrections. It is important to remember that the difference here is often down to how long it will take you to complete the required changes. Naturally, more in-depth changes, such as collecting more data, take longer than correcting some typographical errors. But this decision is sometimes a

Table 6.1 Corrections and awards

Award	What it entails
Ai	No corrections/revisions
Aii	Minor corrections with up to twelve weeks to complete these
Bi	Major corrections with up to six months to complete these
Ci	MPhil. Is awarded
Cii	MPhil. is awarded once minor corrections are made
Ciii	Fail

complex one, based not only on the corrections themselves, but also on how much time you have available to do them. It is useful to remind yourself that major corrections do not necessarily equal a poorer project and also that your corrections are not recorded on your PhD certificate.

When addressing these corrections, it is important that these are addressed in a point-by-point systematic manner, so that the examiner(s) are clear about how each point has been addressed. This could be conducted in a similar way to how you would approach revisions to a journal article. Once these have been approved by the examiners, the graduate school will confirm and you can now be addressed as Doctor!

6.3   Post Viva Celebrations

It is often a long process obtaining a PhD, with many hurdles along the way. Accordingly, after the viva, candidates often experience feelings of exhaustion and numbing. This is due to the exhausting nature of the viva examination; it is a conversation that requires your attention and input for an unprecedented period of time.

It is, however, important you acknowledge this milestone. This can be in either a small or a big way; some may wish to celebrate with peers and others with family and friends. It is important to take a step back and reflect on the achievement. Do so with pride!

Often candidates do not feel that the viva has sunk in until the corrections are complete; however, it is important to realise that as you know in a doctoral journey there are moments worthy of celebration. These should be celebrated regardless of the magnitude of these achievements.

Candidates often comment how post PhD it felt anti-climactic after all the years of working on one project. It can feel like it is all over and candidates often experience a sense of 'loss'. Some candidates express feeling disappointed because they wanted a viva examination that was challenging, and this had not been the case. Furthermore, some candidates feel disappointed at the outcome. All of these feelings are valid and happen to candidates, regardless of discipline or HEI. As you read in Chap. 4, which features individuals' experiences, there a range of outcomes and individuals have experienced a range of emotions following the completion of their vivas.

Summary

- The outcome of your PhD is given to you immediately following completion of your viva.
- There are several possible outcomes of the viva, and prior to your viva you should make sure you are aware of the potential outcomes for your HEI.
- Post viva common candidates experience feelings of exhaustion. This is common after such an achievement.

7

Practice Questions

Stacey Bedwell

Perhaps the most obvious way to prepare for your viva is to practice answering some questions that are likely to come up. Of course, we cannot suggest what your examiners might ask specific to your project or their interests but we can provide you with some suggestions of the general questions that may come up. In my experience, most examiners will ask, in some way or another, for you to offer a brief summary of your project. Not only is this a good question to practice answering for the viva itself, but it's also great practice to explain your research to as many people who will listen. Being able to explain your research, concisely, to non-experts is a key skill when it comes to communicating research, and something that will serve you well, not only in your viva but also when moving forward in your future academic career.

7.1 Common Viva Questions

Every PhD thesis is different; therefore, it is impossible to have a standard set of questions that all examiners ask all students. Your colleague might have completed their viva the week before you and be able to tell you every single question they were asked (kudos to them if they can recall every single question!); this is not necessarily that much use to you if they are specific to their project, however, or if their examiners have different interests to yours.

It is not like being an undergraduate and somehow finding out what's on the exam. However, there are several themes of questions that you can almost guarantee something will come up in. One you know what the examiners are aiming to establish from the viva, you can start to pre-empt what they might be most interested in asking. Especially if you have done your homework on your examiners and know what sparks their interest!

The following pages contain example questions and planned answers for the types of questions that may come up in your viva. The focus of the example answers you can see here are based on a PhD thesis in neuroanatomy; a brief overview of the project is provided below, so you can put the answers in context. In gaining an understanding of the following questions, and how one might answer them, it does not matter if you are familiar with the neurobiology terminology—many answers here should be, to a great extent, understandable to a lay person. For comparison, you will also see example answers from a range of other fields. You will notice that as you progress through this chapter, the example answers provided become fewer. As you work through the questions and continue to attempt your own answers, you should become more confident in answering without examples and prompts. By the time you get to the end, you should be able to go ahead and answer the questions provided without there being any examples to follow.

Whilst it is useful to think about how you might answer these common questions based on your own research, it is important that you do not try to learn or memorise answers verbatim. The purpose of the viva is not to test your memory, but to hold a meaningful discussion of your research between experts. Therefore, it is vital that you can think on the spot and outside of answers you may have prepared.

Many of the example answers given are based on the following project.

Project Title: The anatomical connectivity of the mammalian prefrontal cortex.

Author: Stacey Bedwell.

Field: Neuroanatomy/biosciences.

Studies: Thesis comprised several studies, all using very similar methods to investigate the anatomical connections from an area of the brain called the prefrontal cortex. This part of the brain is associated with high-order functions such as decision-making and forward planning. Each chapter looks at connections from the prefrontal cortex to a different brain region or at varied scales.

Main findings: Connections from the prefrontal cortex to other cortical regions are systematically ordered, much like the rest of the brain. However, there are some key organisational differences that are not seen elsewhere in the brain.

Note that these example answers are suitable for this project. Giving these exact answers to similar questions in your own viva is unlikely to make any sense. However, you can see from the examples the type of information and level of detail it could be useful to provide.

General Questions

To open your viva your examiners are likely to start off with some general questions about your research. This gets the conversation started and will help you settle into the flow of things. It also gives your examiners a good overview of what your project is all about, to ensure you are all on the same page from the start.

Q. *Can you start by summarising your thesis?*

Answer 1:

The project investigated the anatomical organisation of input and output connections from the prefrontal cortex (PFC), in two main pathways in the brain. I used fluorescently tagged neuroanatomical tract tracers injected systematically into the PFC in rats to visualise the connections to these two regions, both on a relatively broad scale and on a much finer scale. Detailed analyses of the location of labels across three axes of orientation revealed properties of the circuits which had previously been either undescribed or not studied in detail. The most prominent and recurring finding across all of my studies is that of the differential organisation of input compared to output connections, showing little evidence of fine-scale reciprocity; that is, that labeled inputs and outputs from the same PFC injection sites were found in different regions. On a broad scale we found that inputs and outputs consistently followed different orders of connections when moving across the PFC. We also revealed interesting evidence of ordering, showing clear differences in the organisation of connections in various regions of the PFC.

Answer 2:

My project aimed to develop an intervention for a decline in sexual health in a specific population. Specifically, it is focused on combatting an increase in diagnosed sexually transmitted diseases amongst people aged 16–20. The project took a mixed-methods approach, to first identify the needs of the target population through qualitative interviews with young men and women as well as healthcare providers. I then used this information to design an education programme and involved stakeholders again at this stage to establish which elements would be most successful.

Students often ask how long they should speak for when asked a generic question like this. The answer is something along the lines of "How long is a piece of string?" Nobody is timing you. Remember it's a conversation; the examiners won't move onto the next question on a list if they are not completely satisfied with the amount of information you have given, and they will likely prompt you for some more detail if they want it. Likewise, if you are going on for what they consider too long, or if they feel you have answered the question but are still talking, they might politely interrupt you and move the conversation on.

Q. *Now, can you summarise your thesis in one sentence?*

Answer 1:
My project investigated the organisation of anatomical connections from the PFC to other brain regions. The most prominent findings were that of differential ordering of input and output connections and a lack of reciprocity.

Answer 2:
My project aimed to combat issues with sexual health in young people, by designing an education intervention based on identified needs.

It is a very useful skill to be able to concisely summarise your research, whether in one sentence or just a few. The three-minute thesis is an excellent way to practice this. Practice summarizing your thesis in a few quick sentences with your family and friends.

Q. *What is the idea that binds your thesis together?*

Answer 1:
Establishing the organisation of PFC connections is the overall aim of the project and a common theme of each chapter. Each chapter looks at a different pathway, a different resolution or a different area of the PFC. Everything comes together to give an overall picture of how connections from the PFC to two other brain regions are organised on a broad and fine scale.

Answer 2:
My project aimed to improve understanding of how jurors make decisions. Each study explored this concept from a different perspective or gathered information in a different circumstance. Combined, the findings come together to help inform how policy could be improved.

It is vital that your thesis comes together to form a cohesive project and that is not a disjointed list of unrelated studies. Asking a question like this helps your examiners confirm how all of your chapters link and what your overall story is. Your answer here demonstrates to your examiners that you know what the bigger picture is and what message you are trying to convey.

Q. How did your research questions emerge?

Answer 1:

The prefrontal cortex is associated with a lot of complex processes which are not fully understood. Much more clearly understood regions, like the visual, motor, and sensory cortex, show that there is often an underlying ordered anatomical map to the functional maps that allow us to understand how a region works. Although there has been knowledge of what regions and structures the PFC is connected to for a long time, the precise neuronal circuitry has remained unclear. My research aims, to determine how the prefrontal cortex is organised, emerged from this need to understand anatomy before function can be fully appreciated.

Have a go at answering the following questions yourself:

Q. What is original about your work?

With this question the examiner is trying to establish that you understand and can articulate what you have added to knowledge. Think about what you did that was different to others: it may be a different methodological approach; it may be applying a method in a new context; or it may be developing a new method.

Q. Which parts of your thesis are you proud of?

This is not a trick question. Here the examiner might be trying to establish your own thoughts about your research and its implications. This is all part of establishing that you did and understanding the work presented in the thesis. Your examiners also are likely genuinely interested in what you think when they ask this; there's no wrong answer. You don't need to be most proud of the part of your thesis that you think will be the most impressive.

Q. Has your view of the research topic changed over the course of your project?

Your answer to this question can give your examiners an insight into how you have developed as an independent researcher over the course of your PhD. You also have an opportunity here to show your understanding of the continued developments in your research field.

Q. *What are the strongest and the weakest parts of your thesis?*

Here the examiner is aiming to establish how well you can recognise the limitations of your work. It is important to be able to defend your decisions but also to be able to take constructive criticism and use that to build future research.

Q. *Where will you publish/disseminate your work?*

At face value this question could appear quite trivial and seem like it will make a difference to your outcome if and where your work is published. The truth is that it should not affect your viva outcome whether you have published papers at that point. It is valuable, however, to have a plan for where and how you plan to disseminate your findings. Communicating research is as important as doing it; having a clear plan shows your examiners you are aware of this and have thought carefully about who needs to know about your research.

Literature Review/Background Questions

To gain a good basis for understanding your work and what it means, as well as establishing your own understanding of where your research fits in the bigger picture, your examiners will likely ask you some questions specifically focussed on the literature and how it led you to the project you carried out.

Q. *What are some of the main issues and debates in this subject area?*

Answer 1:
In research using anatomical tract tracers, one of the main debates is that of the specific properties of some of the tracers commonly used. There is a prominent debate as to whether or not findings truly represent the type of connections reported or not. Fluro-Ruby, one of the tracers I used, was used in my studies as an anterograde tracer, but it is known to have some retrograde properties. In some instances, it has been described as bi-directional. This could have been problematic as I was relying on this labelling to determine anterograde connections. I aimed to combat this debate within my thesis. I conducted some additional experiments comparing Fluoro-Ruby labelling with two other anterograde tracers; my findings confirmed Fluro-Ruby labelling was reliably anterograde. This is not only benefitted my research, but also helps inform future research in this debated area.

Note that the example answer here is focused on a debate surrounding a particular methodology; you could also discuss debates and controversies in terms of findings and understanding from the literature. It's important to remember that the viva discussion can easily be led and directed by you, since you can ensure the discussion goes in the direction you want by carefully considering how you answer questions.

Q. Which are some of the most important publications or papers that relate to your thesis?

Answer 1:

A specific study I cited frequently in my thesis is one that was published towards the end of my PhD project, by Kondo & Witter in 2014. In their paper they reported an ordered arrangement of PFC connections to a specific brain region in rats. This study provided an important basis for my own findings. My findings went on to provide support for theirs, but also add further dimensions and specific structural details of the same pathway they had not identified.

Have a go at answering the following questions yourself:

Q. How does your work compare to other research being conducted in the field?

With this question your examiners are likely trying to establish your understanding of where your research fits into the overall story of the literature. It would be useful here to explain unique factors of your research, that enable you to add to that story. This might be, for example, in terms of methodological approaches, analytical methods, samples studied etc.

Q. What are the most recent major developments in your research area?

This question provides an opportunity for you to demonstrate that your knowledge in your subject area is up to date and that you did not, for example, stop looking at the literature when you finished writing your literature review. In your answer you can also show your examiners that you have an awareness and understanding of how your research fits into the bigger picture.

Chapter or Study-Specific Questions

When it comes to questions specifically about your studies or chapters, this is where the greatest variation between different projects comes in. Questions here tend to be focussed on three main areas: design, analysis and interpretation.

Have a go at answering the following questions yourself:

Q. Design—what would have improved your approach?

It is important here not to list numerous methodological flaws that could have been easily avoided. However, it is fine to consider decisions you might now make differently given hindsight and the more advanced knowledge and understanding you likely have now compared to the beginning of your PhD project. If you made improvements to your method or approach through the course of your project, or between studies, this could be a good time to explain that process.

Q. Analysis—how do you know that your findings are correct?

This could sound like a trick question, but it isn't. If your examiners ask you this what they are trying to get you to do is to justify your decisions and show your in-depth understanding of your research and what it all means.

Q. Interpretation—which of your findings are the most interesting to you and why?

There is no right answer here. What interests you the most could be different to what interests your examiners most, and that is fine. Be prepared to explain why you find a particular finding most interesting; this could be because it has the greatest implication, because it was not what you expected, or simply because it was the part of your project you felt the most passionate about.

Overall Contribution and Value Questions

One of the main requirements of a PhD is to add value to the research field. Your examiners will likely ask you specific questions related to the justification for your research project, why your research questions were important to answer, what your novel findings are, and what the value of your project is both in the specific field and in terms of 'real-world' implications. Your examiners need to establish what new information you have added and why it is useful.

Q. Why is the problem you have tackled worth tackling?

Answer 1:

The PFC remains relatively little understood, in terms of both anatomy and function. There are multiple neurological deficits linked with the PFC, e.g. psychosis, but without a clear understanding of how the PFC works it is not possible to gain a full understanding of these deficits. In addition, the PFC is known to be the newest human brain region and is linked to high-order processes such as executive function and abstract thinking. Developing a clear picture of PFC organisation could, in turn, enable us to develop a clearer understanding of complex processes like these.

Answer 2:

To gain a clear understanding of how healthcare services and interventions can improve the quality of living, develop treatments and prevention for schizophrenia, it is important to gain an understanding of what the lived experience is like, especially in the area of negative symptoms as this is often an aspect of schizophrenia that is overlooked in the literature.

Q. Which of these findings are the most interesting to you? Why?

Answer 1:

The most interesting finding to me is the differences in reciprocity I have described, especially the gradient in terms of reciprocity. This is most interesting to me because it links in with functional observations of increased abstract processing in certain areas of the PFC (in humans), which is also said to follow a gradient. I find it interesting that the PFC appears to be organised differently to other complex brain regions. For a long time, it has been assumed that brain connections need to be reciprocal and ordered in a certain way in order to be efficient and functional; however, these findings point towards something different and more complex in the case of the PFC, which I think may offer an insight into why and how the PFC can perform such complex functions and can integrate information from multiple regions.

Have a go at answering the following questions yourself:

Q. What motivated and inspired you to carry out this research?
Q. Who will be most interested in your work?
Q. What is original about your work?
Q. What, for you, are the most interesting things to come out of your thesis?
Q. How do you envisage your work being used practically?
Q. What have you done that deserves the award of a PhD?

Alternative Approaches Questions

It is important as a good researcher to be able to recognise no research is perfect. Your examiners might ask you some questions specifically aimed to test your ability to recognise the limitations of the approaches you took and to have considered carefully what the alternatives could have achieved. Here you should be able to justify the decisions you made.

Q. What were the alternatives to this methodology?

Answer 1:
Imaging—although this would be useful, it cannot offer such fine detail; therefore, the fine-scale properties I have revealed would not have been seen.

Electrophysiology—This would be useful as a next step, after identifying anatomical organisation. It is necessary to have a understanding of underlying anatomical organisation to get a clear understanding of physiological or functional organisation.

Have a go at answering the following questions yourself:

Q. Looking back, what might you have done differently?
Q. Can you explain why your chosen methodology/approach was the best choice?

Future Research Questions

Questions about future directions are likely to come up at some point. It's important as a researcher that you can see your research and its findings as a contribution to an ongoing cycle, or a never-ending story if you will. If you are able to show a thought-out consideration of where research in your area should or might go next, you can demonstrate to your examiners you have developed this aspect of becoming an independent researcher.

Q. You propose future research. How would you start this?

Answer 1:
The findings from this project give an important insight into anatomical structure. This can go on to inform further investigations into functional structure. Specifically, the knowledge I have gained here about the differences in connectivity in the PFC compared to other regions can inform investigations into the network structure of functional networks. This information can then ultimately lead to clearer understanding how the prefrontal cortex and

associated functions like decision-making develop and can be influenced, for instance by environmental factors in childhood. This can, in turn, lead to clearer understanding of psychological disordered known to involve deficits associated with prefrontal functions, like deficits in decision-making seen in schizophrenia.

Q. How has your view of your research topic changed?

Answer 1:

I set out on this project with the intention of unveiling the detailed organisation of prefrontal cortex connections. The results of my systematic approach soon revealed properties which I had not anticipated. My view before beginning was that the PFC would be reciprocally and topographically organised in terms of connections, just like other regions e.g. visual. However, at the end of the project my view is that the PFC contains both reciprocal and non-reciprocal connections and the alignment of inputs and outputs is likely to be an important contributing factor to how and why the PFC is able to carry out such complex processes. This has also opened up the question as to whether this type of connectivity is unique to the PFC.

Have a go at answering the following questions yourself:

Q. How do you see research developing in the next five years?
Q. Given unlimited resources and funding, what comes next?

Why not try answering some of the example questions yourself? I have also included some additional questions here that you might want to think about. Remember when preparing or practicing answering questions it is important not to memorise your ideal answer verbatim. This exercise should be used as a method to get you thinking about how you will explain things in an accessible way for your examiners, give you practice in how you can articulate complex and abstract ideas and help you identify things that you may have given little consideration.

Additional questions to think about:

Q. What were the crucial research decisions you made?
Q. How did you deal with the ethical implications of your work?
Q. What are the strongest/weakest parts of your work?
Q. To what extent do your contributions generalise?
Q. Who will be most interested in your work?
Q. What is the relevance of your work to other researchers?
Q. What is the relevance of your work to practitioners?
Q. Which aspects of your work do you intend to publish—and where?
Q. What are the contributions to knowledge of your thesis?
Q. How long-term are these contributions?
Q. What are the main achievements of your research?
Q. What have you learned from the process of doing your PhD?
Q. What advice would you give to a research student entering this area?

Summary

- You cannot plan for every question that will come up; this is the nature of the viva. It is a dyadic (two-way) conversation.
- Whilst it is a good idea to think about how you might answer and explain the points that are likely to come up, it will not be useful to try to memorise answers.
- There is no right or wrong answer to any question. Your examiners want to hear your independent thoughts and engage in an academic discussion.

For reference, the main thesis discussed in this chapter was submitted by Stacey Bedwell in partial fulfilment of the requirements of Nottingham Trent University for the degree of Doctor in Philosophy in April 2015.

8

Being Confident in your Thesis

Isabelle Butcher

It is often the case in academia that we focus on what lies ahead, the next goal, the next hurdle that must be overcome. The 'danger' of this is that we do not acknowledge our achievements. It is worth stopping after submission of your doctoral thesis to acknowledge the milestone you have achieved. Pausing allows you to reflect on the work that you have conducted and the achievements that you have made to date. It is crucial to acknowledge these inchstones as well as the more major milestones in a doctoral journey.

8.1 How to Ensure you Are Confident in your Work

1. Reflect on what your specific doctoral work adds to the field. How does your study add to the existing knowledge already known about a particular area? After spending so many years on a PhD it can be challenging to see the greater picture of how your research fits into this. Yet it is crucial that you consider this, not only does this increase your confidence in your own doctoral work; it also gives you an insight into the importance of your research to the wider research field.
2. Make opportunities to share your doctoral work with others both within your research field and outside of your immediate research field. So often

as researchers we share our findings with those who work with us rather than considering that many of our findings are transferable to other disciplines. Speak at conferences and talks that are not in your immediate research field, speaking to those who are not experts in your area will highlight to you the crucial components of your doctoral work and those aspects that are applicable to other disciplines.

3. Consider your doctoral work step by step, break down the thesis from the broader question into the smaller research question(s) and aim(s) and consider how these all fit together, and be clear about how aspect of the doctoral work answers the wider research question/title of your PhD thesis.

4. Reach out to non-HEIs, for example charitable organisations and organisations that will not be familiar with academic presentations. Presenting to a range of audiences enables individuals to question different aspects of your doctoral work. This is an important skill that every researcher must be able and feel confident in doing.

5. Make a list of all those inchstones you have achieved in your doctoral journey; consider how these have impacted your journey and work. Inchstones can be any level of achievement: for some candidates, this may be meeting a recruitment target for an empirical study; for other candidates, this may be having a paper submitted in an academic peer-reviewed journal.

8.2   Your Work Is Valuable

It is important that you recognise that, regardless of the magnitude of the work, it is valuable. Whatever the size of the study, your work is novel and has not been conducted in the way that you did it before. It is important that you realise this; this realisation impacts the way that you approach your viva examination. Those candidates who acknowledge the value of their doctoral work are often more articulate when considering the 'real-life' implications of their work. One way of considering the value of your work is by considering how it fits into the research picture. For some candidates, this can be challenging if, for example, there may be no apparent 'real-life' implications for several years, but there will always be 'real-life' implications to every doctoral work undertaken. Some candidates choose to undertake science outreach work with school-aged children; this gives individuals a sense of the wider picture and the importance of their doctoral research to wider society.

8.3   You Are the Expert

Candidates often fail to recall that they are the experts in their thesis. Your examiners may have examined far more theses and have published more papers, but they are not an expert in your specific doctoral work. It is therefore important that you recognise this and keep this in your mind as you prepare for the viva as well as when you are in the viva. Some candidates can feel that the examiners are asking 'simple' questions to trick them, but as candidates you need to remember that the examiners do not know your work to the extent that you do. You should be prepared in the viva to justify reasons and to clarify why you did what you did. The examiners do not know the literature in your doctoral work to the extent that you do.

8.4   Imposter Syndrome

This phenomenon is not new; it was first identified in the 1970s (Clance & Imes, 1978). It can be defined as doubting your abilities and feeling a fraud. Labelling these feelings imposter syndrome is controversial, with some suggesting that the label 'imposter' suggests a criminal fraud, and the medical term 'syndrome' suggesting that this is a medical condition. But it is not. These feelings are a normal part of the professional world, which we can all attest to have experienced at some point. Particularly unhelpful is the concept that individuals 'suffer' with imposter syndrome, rather than simply 'experiencing' these feelings. It is common for doctoral candidates to express doubt over their work; as a doctoral candidate by reflecting on the work that you have successfully executed this should serve as a reminder that both you and your work are valuable. Here are some approaches to overcoming these feelings:

- Recognise the benefits of not knowing everything. Embrace being a novice. If you do not know the answer to a question in the viva, do not panic, consider it and answer it in a way that opens up the discussion. You are then engaging in an intellectual discussion.
- Focus on how much you have learnt within your doctoral studies rather than how much you have done. Do not be concerned with how many studies you have conducted or how many papers have been published. Focus instead on a growth learning mindset. Foster this.

– Be aware that others, whether they are senior researchers or your peers, may experience the same feelings as you. These feelings are normal.

At this point it is important to reiterate that you are not your PhD. Your doctoral work does not define you.

8.5 Defending your Thesis

It can be a difficult when, as part of your viva, you are asked to defend why you did something in the way that you did it. Students can become defensive, but the risk of becoming defensive is that this closes the door for intellectual debate and conversation. One way in which this can be prevented is by stating to the examiners why you did what you did in the way that you did it, and acknowledging you could have done it differently. Explain why chose not to do it in any other way and be clear about the reasons. The examiners are not trying to trick you; they are genuinely showing an interest in your work and they need to be able to understand the decision processes behind your work.

8.6 Saying 'I' Not 'we'

In academia 'we' is often used because much of the work that is conducted and researched is a collaborative effort and therefore 'we' is correct. In a research study, for example, there are many people involved, from those collecting the data, to those analysing the data, to those overseeing the running of the study. It is important to acknowledge the work each team member plays in the work that is being conducted.

In your viva examination it is not the place to say 'we' because whilst some decisions may have been made with your supervisory team or through discussions with your research group, ultimately you alone are responsible for the decision as it is *your* PhD. When speaking about your doctoral work it is essential that you state 'I' when referring to aspects of your work, since this helps you to 'own' the thesis and take full ownership of it. Rather than saying 'we' if you would like to acknowledge collaborative decisions made during your doctoral work that had an impact on your theses, be clear and say 'in collaboration with my supervisory team and research group'. If you continually say 'we' in your viva the examiners will be unaware of firstly who 'we' is and "f" for all doctoral candidates is the ability to be able to reflect on the limitations of your work, whilst not undermining the work that you have completed. Often doctoral candidates fall into the trap of either failing to acknowledge apparent limitations of their work or acknowledging the limitations to such an extent that questions why they conducted their doctoral work in the way that they did.

8.7 Honesty and Hindsight

As doctoral candidates you are aware of the limitations or weaknesses of the doctoral work you have conducted, but it is important that you are able to articulate these in the oral viva examination. Examiners may not ask you directly what the weaknesses are, but they will often ask for justification of why a doctoral candidate conducted something in the way that they did. Within the justification it is expected that you, as a doctoral candidate, will be able to draw out possible weaknesses.

It is expected that these will be written into the thesis, yet the examiners need to be sure and know that you are aware of these limitations. It is not sufficient to know the limitations, however. It is important as a researcher that you are able to state why you did what you did, and why this is a limitation. This is one aspect of the oral examination that you can indeed prepare for; it will happen through a thorough consideration of your doctoral work and an untangling of the reasons behind each of your actions. For those who publish their doctoral work and present it in a thesis by journal format, where each chapter is a paper, the weaknesses will have been clearly stipulated for each study.

When sharing limitations of your doctoral work it is, however, crucial that you do not undersell your work, and that do not blame external forces for the weaknesses. Some doctoral candidates blame the data—that it did not do what they wanted—whilst others dismiss possible weaknesses.

When sharing the weaknesses of your viva in your oral examination it is vital that you are able to show that, even with the weakness(es) with your study(ies), your PhD achieved what you aimed to. It is crucial that all weaknesses are considered in light of the overall research question(s) of your PhD. It is also important to be open about why a decision was made, despite there perhaps being some known limitations of this at the outset.

Furthermore, in a viva you should be able to articulate and consider what you might have done differently if you could do the work again with the knowledge of hindsight. It has undoubtedly been a short time since you conducted your work, but the power of hindsight is such that as soon as individuals have completed a task, 'in hindsight' is an experience that is rapidly experienced. It is entirely acceptable and, in fact, it is welcomed in your viva to state 'in hindsight…'. The crucial aspect to this is that you can consider how you may have completed a task differently as well as articulating why you would do it differently and how completing this in a different method would impact the results of your PhD. This ability to reflect on one's work is important and is an experience that all individuals experience regardless of the career and life they lead. In a viva examination, it shows that the doctoral candidate can think independently and engage in conversation that is not scripted.

Summary

- Be confident in your doctoral work.
- Use 'I' in the viva examination not 'we'.
- Clarify why you did something by being explicit, but acknowledge there are alternative methodologies or similar that could have been utilised.

Reference

Clance, P. R., & Imes, S. A. (1978). The imposter phenomenon in high achieving women: Dynamics and therapeutic intervention. *Psychotherapy: Theory, Research & Practice, 15*(3), 241.

9

The Viva Preparation Timeline

Isabelle Butcher

Doctoral candidates ask when they should start preparing and where to start. It can be overwhelming after writing a thesis to look back at the document and approach viva preparation in a structured, objective manner. Some candidates state that they do not wish to look at the thesis again, whilst others look at every sentence and struggle to read the thesis through without seeing typographical errors. It is important to strike a balance.

In this book we have touched on some key viva preparation skills, such as knowing what to expect, being prepared for specific questions and being able to reflect on your doctoral work.

1. The first task to be conducted is to go through your thesis in a systematic method placing sticky notes at the start of each chapter, so in the viva you can clearly turn to these pages. This gives you a clear task and means that you do not read the thesis as if you are reading a book. Reading the thesis as a book, cover to cover, is not helpful once it has been submitted.
2. Be aware of who your examiners are—it is expected you will be aware of who these will be prior to thesis submission.
3. Confirm the date of the viva—often the specific date is not firmly set until nearer the viva due to examiners' work schedules, but you will usually have an approximate date, such as the month in which the viva will take place, when you hand in your thesis.

© The Author(s), under exclusive license to Springer Nature Switzerland AG 2022
S. Bedwell, I. Butcher, *How to Excel in Your Doctoral Viva*,
https://doi.org/10.1007/978-3-031-10172-4_9

4. Go through the thesis with Rudyard Kipling's six questions in mind—
WHAT, WHY, WHEN, HOW, WHERE and WHO (see Table 9.1).
These will enable you to see the bigger picture of your doctoral work.
These questions are as follows. You need to able to answer these questions
for each aspect of your doctoral work. If you have conducted four studies
as part of your PhD then you need to be able to answer each of the six
questions for each component of the these.

> I keep six honest serving-men
> (They taught me all I knew);
> Their names are What and Why and When And How and Where and Who.
> I send them over land and sea
> I send them east and west;
> But after they have worked for me
> I give them all a rest.
> I let them rest from nine till five, For I am busy then,
> As well as breakfast, lunch, and tea, For they are hungry men.
> But different folk have different views; I know a person small—
> She keeps ten million serving-men, Who get no rest at all!
> She sends 'em abroad on her own affairs, From the second she opens her eyes
> One million Hows, two million Wheres,
> And seven million Whys!
> (Kipling, 1900)

5. If any questions come to mind that you are dreading or that you do not
want to be asked, it is important that you address these. Viva preparation
is the time to ask yourself how you would answer these. You can, of course,
seek advice from your supervisory team and peers but remember that only
you conducted your PhD. It is important that these questions are not
ignored in viva preparation. One key aspect of preparation is that you feel
confident in being able to answer any questions that the examiners ask.

Table 9.1 Kipling's six questions

What	*What you did*
Why	*Why you did it*
When	*When you did it*
How	*How you did it*
Where	*Where you did it*
Who	*Who you did it with*

6. In preparation of the viva be clear on the research that your examiners have conducted and published. This may not be cited in your thesis, as it may not be directly related to your work, but it is important that you understand their work, as this will help you to understand their perspective on a topic/aspect of research. This can include a quick literature search prior to viva and does not need to be a long systematic search.

9.1   Three Months before Oral Examination

- It is expected that you would have submitted your thesis. This may depend on your HEI and your circumstances.
- Be aware of your viva date and the location, and the practical aspects of your viva such as: When will it take place? In which room? Has the doctoral college confirmed this? Have the examiners been sent the thesis?
- Find out the names of your examiners.

9.2   Two Months Prior to your Viva

- Re-read your thesis as stipulated in the points above. Be clear and structured in your reading of the thesis, chapter by chapter. Reading the thesis as a book is not helpful as it can lead to candidates 'skimming' aspects of the thesis because they assume they know it. Whilst they do know it, this can lead to candidates not engaging with the thesis in a constructive way. At this stage candidates do need to be confident in pulling the thesis apart and spotting the limitations and strengths.
- Highlight aspects of the thesis on which you perhaps need to focus your preparation: for some candidates, this may be the methods sections; for other candidates, this may another aspect. This action of highlighting ensures that you do not miss out on any of these aspects. Often candidates do not want to acknowledge the trickier aspects of their theses; by not reflecting on these, however, this can lead to students stumbling in the viva. Now is the time to be questioning your decisions, since you have a safe environment and both the resources and the time.
- Address any concerns or queries about your doctoral work with the supervisory team. For some, this may be as part of a mock viva—which has been addressed elsewhere in this book. For others, this will involve a more collaborative discussion. It is important that you utilise your supervisor's expertise. Whilst they are not experts in your doctoral work, they can offer

solutions or guidance as to how they approach a specific question that may arise.

- Make the most of opportunities to share your research; for example, now is the time to give talks to charitable organisations, other HEIs, other research groups. Through speaking about your work, you are consolidating your knowledge of your doctoral work which is critical to your performance in your viva.
- Consider, and reflect upon, the story of your thesis. Often PhD questions have one broad research question and aim and candidates have, in addition, several subsidiary research questions and studies. This is acceptable but it can occur that candidates are unable to join up the studies and stipulate how each aspect of the doctoral work answers the PhD question(s)/aim(s). Failure to tell the story of your thesis in a clear and succinct manner can make it challenging for the examiners and other individuals to see how your research fits together. PhDs are often akin to jigsaw puzzles, with one central research question/aim. To complete the jigsaw, there are several elements, studies which together with each study's results enables the linking of the pieces.
- Consider how you would draw your PhD. If you could draw your PhD, what would feature? Whilst it is not necessary to draw the PhD in the viva, for some candidates this is a usual way of visualising the PhD and showing how elements of the doctoral work links together.
- Look at questions. We have addressed this in a previous chapter; it is important to consider questions, but it is also crucial that you do not simply memorise questions and answers. This is not helpful and is not what the examiners are looking for. The examiners need to see that you can think independently on the spot; of course, some questions are predictable, such as 'Why did you do this PhD?'

9.3 One Week to Go

- With your family and friends, practice the short summary of your thesis from beginning to end.
- Be clear about the location of the viva, who is present and the timings involved.
- Consider what you will wear. It is important that you are smart but also comfortable.

- Ensure that you can remain focused on preparing for the viva without having other competing interests on your time.
- Remember to rest and sleep well.

9.4 The Day before

- Candidates have likened the day before their viva to the day before a job that they really want, and it is common to experience feelings of jitters, and restlessness. These feelings are all normal and it is best to remain calm and reflect on all that you have achieved to date.
- Now is not the time to be reading papers around your thesis. Instead, remember and reflect on what you know.

9.5 The Viva Day

- Eat well.
- Leave time to travel to venue(s) and allow extra time.
- Ensure you have a copy of your thesis to take with you to take into the examination.
- Remember all that you have achieved to date—you have successfully submitted a doctoral thesis. That is an achievement in itself!
- Be confident in your work.

9.6 Preparation Timeline

This timeline will help guide you in preparing for the viva examination. Make it as structured or as fluid as you like, depending on your needs.

Viva timeline worksheet

<u>Expected date of thesis submission:</u>

<u>Expected date of Viva:</u>

<u>Date to start preparing:</u>

<u>Number of weeks of preparation:</u>

<u>Day of viva</u>
Main goal:

<u>Day before viva</u>
Main goal:

<u>Three days before viva</u>
Main goal:

<u>One week before viva</u>
Main goal:

<u>Two weeks before viva</u>
Main goal:

<u>Three weeks before viva</u>
Main goal:

<u>Four weeks before viva</u>
Main goal:

Add in additional weeks here if you need to.

Summary
- It is important to consider how you plan to prepare for your viva examination.
- When preparing for your viva make sure you factor in a break from your thesis so that they you can look at the thesis with 'fresh eyes' post thesis submission.

Reference

Kipling, R. (1900). I keep six honest serving men. *Just So stories*.

10

Participating in a Mock Viva

Stacey Bedwell

A mock viva is essentially a practice run. They take many different forms, with some being quite serious events whilst others are much more casual affairs. Mock exams in some shape or form are something with which most of us are very familiar; most exams in the UK education system are proceeded by a mock. They have their uses in familiarising the student with the format and exam procedures. They also arguably have their disadvantages. Not every PhD student has or feels the need for a mock viva. Some institutions require you to do one, whereas others do not. Whether or not you find a mock viva useful is largely going to be dependent on what you want to get out of it.

10.1 Why Do a Mock Viva?

Whether or not a mock viva is a useful process is quite dependent on the individual and their circumstances. From reading the personal accounts of viva experiences in Chap. 4, you will have recognised that some PhD candidates found the mock viva very beneficial and would strongly recommend one. Others did not find it useful at all, or sometimes did not find it useful in the context of a mock viva, but simply say that it was a beneficial conversation. With this in mind, answering the question as to why you should do a mock viva can be tricky. You should participate in a mock viva if you feel it

© The Author(s), under exclusive license to Springer Nature Switzerland AG 2022
S. Bedwell, I. Butcher, *How to Excel in Your Doctoral Viva*,
https://doi.org/10.1007/978-3-031-10172-4_10

will be beneficial to you. Some institutions require a mock viva as part of the PhD process, and the student does not get a say in such cases.

If it works for you, and you plan it well, with the right people, a mock viva can do a great job at removing some of the unknown element of the viva itself. If your mock viva is set up as close to possible to what the real one will be like, you will be far more familiar with the process on the day and not going in as blind as you might do otherwise. Of course, this could backfire, if your mock viva were to go badly, this could end up inducing more anxiety than necessary. This is why, if you do participate in a mock viva, you should plan it well.

10.2   How to Set up a Mock Viva

Once you have made the decision to have a mock viva, you need to ensure it is as beneficial as possible to you. This means thinking about how and when to set up the mock viva and what it should include.

First you should decide if you will find it helpful to set up your mock viva to resemble your real viva as much as possible. This might depend on the reasoning behind your mock viva in the first instance. For example, are you having a mock viva primarily to practice answering questions and articulating your research, or is it more to remove the feeling of the unknown surrounding the viva itself? If the latter, you should consider emulating the real thing as closely as possible. This means that if your viva exam is going to be on campus in a meeting room, you should also hold your mock viva on campus in a meeting room. It would not be a very accurate emulation in this regard if you held your mock viva via video conferencing whilst sat in the familiar and comfortable surroundings of your own home. Similarly, virtual meeting dynamics and the way we interact with one another in an online meeting is very different than in person. Therefore, if your viva exam is planned to be held online, I would recommend you try to hold the mock online too.

When it comes to your mock viva, you naturally have more control over what you would like to be covered. Whilst it is not a great idea to avoid topics or questions that make you uncomfortable, the mock viva is a good opportunity to ask your mock examiners to ask you about specific things. These should perhaps include topics you feel you need to practice talking about or on which you want some feedback. If time is limited in the mock viva, which, in reality, is often due to the busy schedules of academics, it may be a good idea to think in advance about two specific areas about which you feel most confident.

Don't waste precious time covering these topics when you could be utilising the opportunity to focus on areas you find more difficult.

To get the most out of the mock viva is it important to do a little preparatory work and organisation. Ahead of the mock viva, spend some time communicating with the mock examiners about what you would like to get out of the process. It's also important that you give your mock examiners enough time to read your thesis. Granted they will likely not have as much time to do this as your actual examiners, so they won't be able to go into as much detail, but you do need them to have had a chance to look at it in order to be able to ask you questions specific to your project. If time commitment is a concern, ask each mock examiner to focus on different chapters in order to reduce the load. For example, one examiner might cover the introduction and two experimental chapters, and the other might cover two different experimental chapters and the discussion. You could even ask them to focus more on specific chapters you are more concerned about and to ignore those about which you feel more confident. This is your mock viva, so try to tailor it to your needs as best you can.

After the mock viva try to schedule some time immediately afterwards to discuss how it went with the mock examiners. This way everything that you discussed is fresh in everyone's mind; it's just as important that you remember as them. Ask your mock examiners to provide you with some written feedback notes and take some notes yourself. At this point it is useful if you have given your mock examiners an idea of the kind of things you are most worried about beforehand, so they know what you specifically would like feedback on. Encourage them to give feedback outside of your pointers, however; they will likely pick up on some areas for improvement that you wouldn't. Perhaps the most useful feedback you can get from the mock viva process is more feedforward; you should ideally discuss aspects of the mock viva that you can improve before the viva. It would not be helpful at this point, for instance, for your mock examiners to tell you that the method you used was the wrong one, suggesting that you use another one, since you cannot do anything about that now. However, it could be useful to discuss possible alternative methods, to ensure you can justify the methodological choices you made and to explain clearly why you think this was the best approach. Similarly, whilst it would not be helpful for your examiner to suggest you rewrite a whole chapter or run a completely new analysis if your viva is just weeks away, it could be helpful to know where clearer explanation might be needed or where there might be room for future research.

Having gained useful feedback, or feedforward comments that you can act on, it's essential that you do action them. This, after all, was the purpose of the mock viva. For most, the mock viva will take place after the submission of the thesis. Therefore, action points from your mock viva will be quite different to the corrections you might receive from your real viva; you're not about to go changing your thesis content at this point. Mock viva action points are likely to be more focussed on things like articulating things you may have not explained very clearly, ensuring you can justify all decisions and generally working on being able to confidently discuss the work you have done.

10.3 How Far in Advance Should the Mock Viva Be?

The timing of the mock viva is very important. It's vital that you don't hold the mock viva too close to the real viva. If something were to come up in the mock that you were unprepared for, you need to give yourself time to work on this. Additionally, it's no use having the mock too soon. It's important you have given yourself time to prepare properly beforehand, to get the most useful feedback possible. Having said this, there is no standard number of days or weeks ahead of the viva date when it is ideal for everyone to hold their mock viva.

When is the most appropriate time could depend on how much time you have to devote to viva preparation and what you aim to get out of the mock viva process. For instance, if you are working in a full-time post doc job with no time during the work-day to allocate to viva preparation, to be able to utilise feedback from your mock viva you likely want to hold it at least a few weeks in advance, because you might only be thinking about your viva in depth on weekends. Whereas if you are not working elsewhere and can theoretically devote every day to your viva preparation, just one or two weeks in advance is likely more than enough. In deciding when your mock viva should be, it is helpful to think more widely and consider your whole viva preparation timeline (see Chap. 9).

10.4 Who Should Examine the Mock Viva?

Now you have decided a mock viva is a good way forward for you, it's important to take them time to carefully consider who you want to be involved. Your default thought could your supervisory team, although this might make immediate sense, they are not necessarily the most helpful choice. Consider carefully what you want to get out of the mock viva process and let this inform your choice.

Your Director of Studies/Primary Supervisor

After yourself, your director of studies or primary supervisor is probably the person most involved in your research. For this reason, it might feel like they are the natural choice to be involved in your mock viva. After all, they know the subject matter well, they know what you did and why, they are likely an expert in the field, and they will know relevant questions to ask—perhaps even without having to spend hours reading your thesis verbatim. Whilst all these reasons are probably true, they may actually make your director of studies a very unhelpful choice. Whether or not this is true for you depends on what you aim to get out of the mock viva. As an integral part of your supervisory team, and therefore the research time, your director of studies may be very closely involved in your research by nature. Some far more than others, and this depends a lot on your project and your relationship, but they are involved. As your director of studies, they know your project inside out. They have seen it develop and evolve from the beginning proposal. They know which parts you struggled more with and which parts you are most passionate about. They probably already know which findings you wish were different, and which experiments you would do differently if you could, and they already know why you interpreted the results the way you did. Whilst all of these properties are great in helping you prepare for your viva, perhaps through an in-depth conversation, they can also prevent your director of studies from being able to take the perspective of an outsider, and represent your examiner, when it comes to a mock viva. In your real viva, your examiners have not worked on your project, and they will come to the table with a different perspective and different experiences. The best way to 'mock' your viva as accurately as possible is to aim to emulate this. If what you hope to get from your mock viva is more of a conversation, and if you think that detailed knowledge of your project's evolution is helpful in this case, then, of course, your director of studies is a fantastic choice.

Other Members of your Supervisory Team

The relationship PhD candidates have with their wider supervisory team can vary greatly. For some, they are all equally involved in the project and for others the second and third supervisors may barely be involved at all, or perhaps just focussed on one specific element of the project. When deciding if your wider supervisory team should examine your mock viva, you might find it useful to consider how involved they are. For instance, if you have decided you want examiners that do not know your project inside out, but have enough expertise to ask in-depth questions without too much extensive preparation on their part, a second or third supervisor that knows the project but who has not been heavily involved on a weekly basis could be ideal. On the other hand, if you want a mock examiner who is a stranger to the project, any member of your supervisory team is probably not going to be the right fit.

There are some advantages to involving supervisors who know the project, but not in too much fine detail. Those who have been involved in your project, albeit from a little more distance than perhaps your director of studies, will have a basic understanding of what your research is all about. They will also likely have some knowledge as to where your strengths and weaknesses lie, meaning they can tailor the mock viva more towards building on your weaknesses. Your wider supervisory team are likely to be part of the project because they have expertise in some aspect of your research, this means they have the background knowledge and understanding to be able to ask specific questions that a non-expert might not be able to.

Depending on the dynamics of your team and your own circumstances, you may have a less close relationship with your second and third supervisors than you do with your director of studies. If this is the case for you, this makes the interaction a little close to the real thing than a mock viva with your director of studies might be.

Faculty Involved in Previous Steps

Whether those who have been involved in your progress reports, transfer, upgrade, continuation viva and so on are allowed to be part of your viva is very institution-dependent. For many, the university stipulates those examiners involved at earlier stages cannot examine the final thesis and viva. This could make those people ideal for your mock viva. Anyone who has examined any stage of your project already has a reasonable understanding of your project, they may also have good knowledge of anything you struggle with and can focus on those areas which are of the greatest concern to you. Discussing these areas of your work in detail at the mock stage can help you overcome any specific anxieties you have about them. It is also of note that you likely selected any previous examiners based on their expertise in the research area, so they will be a good fit here too.

It is worth noting that involvement in previous stages could be a negative when it comes to the mock viva, especially if you found those previous stages very challenging and perhaps and have negative associations with that specific person and a viva situation. This doesn't necessarily reflect badly on the individual; it is just a correlation you have made from your own experiences. It's important to remember that the viva is about the project that is presented in the thesis. Depending on your experience, there is potential that any previous examiners have knowledge of struggles and hurdles that your real viva examiners would not. Whether you feel this is a barrier or an advantage to your mock viva depends on your own personal feelings.

Other Experts

Other experts could mean a range of different people. An expert in your research area could be a faculty member who is not part of your supervisory team and has not been involved in your project so far. This could be, for example, someone else in your research group, laboratory, or department. It may mean an expert from another university with whom you have a connection. The important points are that they are an expert in some aspect of the project you have done, and that they are not involved in your actual viva. These people make great mock viva examiners because they are far enough removed from your project to be void of any preconceived ideas, but they also have the knowledge and understanding to be able to discuss and question your work in depth.

If planning a mock viva very far in advance of your viva, perhaps ahead of appointing examiners, or if you want to try one at the transfer/upgrade stage, be careful not to involve anyone who you might like to examine your real viva to avoid further complications.

Non-experts

At first thought you may dismiss those who are not experts in your research from being involved in your mock viva. However, non-experts should not be dismissed without serious consideration. Whether or not a non-expert makes a suitable mock viva examiner depends on what exactly you are looking to achieve from the mock viva process. If you are seeking in-depth, topic-specific discussions, then someone with no prior knowledge of the subject is probably going to be of little help. Similarly, if you are hoping to discuss your novel and complex methodology in detail, then someone who has only ever worked theoretically or in an entirely different methodology might not be the best choice. However, for those methodology discussions, the most appropriate person for your mock viva may well be a researcher who uses the same method in a different field. They are not an expert in your research area, but they will be able to offer a detailed discussion of the method you have applied.

10.5   Should Anyone Else Attend your Mock Viva?

When it comes to your real viva there may be people other than yourself and your two examiners in the room. Commonly, an independent chair is also present. You may also choose to have your supervisor present, although be aware they cannot say anything or help you in any way. When it comes to the mock viva, the structure is generally under your control, so you could theoretically invite along as many people as you like.

Supervisors

Whether or not you plan to have your supervisor in your actual viva, it is worth considering the pros and cons of asking them to attend your mock viva, assuming they are not playing the role of examiner. When it comes to useful feedback, it could be beneficial to have at least one of your supervisors present. This could be especially useful if you have taken the route of having non-experts as your mock viva examiners. Your supervisors, specifically your director of studies, probably knows your project in quite some detail, as well as knowing you well as an individual. They will be in a good position to take an observant role and to give you feedback on how you answered questions, how well you articulated complex ideas and how well you defended your research. On the other hand, you may feel that an extra body in the room is unnecessary if your mock examiners can give the same level of feedback, in the context you require. Some institutions might not allow your supervisors to attend, so you should check this with your own doctoral college.

When it comes to having your supervisor in the room, many students find it off-putting, in both the real viva and the mock. For some, it can increase the feeling of pressure and add anxiety to an already quite demanding situation.

For others, they are somewhat comforted by the presence of their supervisor. This again is a decision that comes down to you and your relationship with your supervisors.

Audience

It is not common in the UK to have an audience in the viva exam, but it is common practice in many countries. In some places, the viva is a defence to an audience, much in the format of a lecture or a conference presentation, where the audience members ask questions. Although not the case for the viva exam in the UK, it is often a requirement within the three years of a PhD programme to present to an audience, for example at the transfer stage or at a school conference. Presenting research to an audience and engaging with that audience is an important skill to develop as an independent researcher, for which you are training to become as a PhD candidate. For this reason, it could be beneficial to include an audience of some form at your mock viva, especially if you have had limited chances to engage with audiences before this point. Including an audience in your mock viva won't necessarily provide much benefit to staging an accurate representation of what your real viva exam will be like, but it would be beneficial to your overall development as an academic. Whether you want an audience could depend on what you feel you need to develop and what you hope to get out of the mock viva process. Arguably, if you invite a full audience along it stops being a mock viva, as we know them in the UK, and moves more towards becoming a presentation of your work.

10.6   What to Do after the Mock Viva

What you do after the mock viva is equally as important as the mock viva itself, or discussion with your supervisory team if you prefer to forgo to the structured mock viva and go down that route. The whole process could be wasted if you don't properly utilise any feedback you receive. The main purpose of the exercise, along with removing the element of unknown from the viva experience, is to build your viva skills and identify anything you need to work on before the real thing. For this reason, it is vital you communicate with your mock viva examiners or your supervisors beforehand, so they know what you need to get out of it.

Get as much constructive feedback as possible out of your mock examiners. This is a time when being reticent about things that need improvement will be unhelpful. If you explain a concept unclearly or in an inaccessible way, you need to be told now, so you can work on improving your approach and articulation for your viva. It could be helpful to pinpoint some areas you are more concerned about and highlight these to your mock examiner, so you can receive specific feedback on these points. Structured and clear feedback, with useful feedforward comments, can be put to good use in the days and/or weeks between now and your real viva.

After you have completed the mock, you might find it useful to step back from the thesis for a couple of days before revisiting the feedback you received and thinking about an action plan moving forward.

10.7   Key Elements and Questions to Include

The exact questions covered in your mock viva are going to be very dependent on your project and what you are struggling with most. However, there are some key elements I recommend you ensure are included. It would be useful to discuss these with your mock examiners beforehand.

- A summary/overview of your whole project.
- An explanation of the need/inspiration for your project.
- A summary of each study/chapter/experiment.
- An explanation of how each study links or theme that ties all chapters together.
- Clear explanation of the main findings and their value.
- An idea of what the future implications are.
- Recognition of limitations.

One of the purposes of your mock viva is to highlight anything you struggle with or find difficult to articulate. Ensure your examiners can give you critical feedback on this, with useful feedforward comments so you know what you can do to improve.

If you are able to easily identify key areas or questions you are keen on covering in detail in the mock viva, this could inform who you choose to act as a mock examiner, depending upon their expertise.

Summary

- Whether you should have a structured mock viva, who should examine your mock viva and what it should include depends on what you aim to get out of the process.
- It is often helpful to seek out academics outside your supervisory team to conduct your mock viva.
- If you have a mock viva, ensure you get the most out of it as possible. Talk to your mock examiners beforehand, so they know what you are most concerned about.

11

Viva Concerns

Isabelle Butcher

A PhD is likened to running a marathon at sprint speed. There is little time to stop and look, but the time between thesis submission and viva examination is a time where it is crucial that one takes the time to stop and pause before the final hurdle.

A PhD is a long journey with many ups and downs, and for many candidates the troughs often seem more memorable than the peaks. It is a journey where you not only discover more about your specific doctoral area of work, but you also discover more about yourself. You understand how you work best—with deadlines or with flexibility, at night or in the early mornings. You learn how to work alongside people and you also understand the importance of independent learning and thinking.

One way in which viva concerns can be dealt with is by ensuring that you remain well and that you take care of yourself in the lead-up to the viva. It is important that you look after yourself and your own well-being. This term obviously means different things to each of us, and at the heart of well-being is understanding what it means to you? (Butcher et al., 2022). For some people, this is being alone in nature; for others, this is being with family and friends. To ensure you remain well for the viva it is important you make time for those activities that are good for your well-being.

There are several ways that you can look after yourself, in particular using the four 'R's': *remember, reflect, refresh and rest.*

S. Bedwell, I. Butcher, *How to Excel in Your Doctoral Viva*, https://doi.org/10.1007/978-3-031-10172-4_11

I. Remember. Remember all that you have achieved to date. You have successfully completed a thesis; that is a milestone that is worth remembering. Remember all those obstacles that you have overcome in your PhD. Each candidate experiences obstacles in their doctoral journey, but it is important to reflect and remember those occasions that you have overcome obstacles. Remember these by writing them down and processing the impact of this event, how you felt at the time and, looking back, ask yourself: How did that experience impact on the rest of your doctoral journey? For many doctoral students, recruiting participants for studies is an obstacle which can seem unsurmountable; however, once these candidates have overcome this obstacle and met their recruitment targets(s) they can look back with pride of their achievement.

II. Reflect. This is the time to pause and reflect on events that occurred along the doctoral journey, which perhaps did not go as planned. Consider, through reflection, how you may approach a future situation differently knowing what you know now. Reflect on all you have learnt during the doctoral journey, all the skills acquired and all the people you have had the opportunity to meet, whether that be participants, colleagues or collaborators.

III. Refresh. Take time to step away from the thesis and refresh your mind. Spending a long period of time on one piece of work can be exhausting, both mentally and physically. It is important that you seek time away from your thesis before the viva. Refreshment is different to each person, for some, this may be going away for a change of scenery; for others, this may be stopping and being still in their current environment. Candidates often state that they need to look at the thesis every day between submission and viva examination; this is simply not the case. It is important that you are refreshed so that you are looking at your thesis with fresh eyes. If not refreshed, you can miss possible points in your thesis and thus it can hinder the outcome that you receive.

IV. Rest. This is an activity that should occur in the course of viva preparation. During viva preparation, it is important to take moments to rest in a way that works for you. Do not compare your idea of rest with that of other people.

Whilst there are a host of activities that you can do to maintain good well-being, some of the key activities which are known to help include, but are not limited to:

- Green spaces—get out in nature, whether that be your garden or a local park. Evidence has consistently reported that being in green spaces increases an individual's well-being (Ajayi & Amole, 2022; Peacock et al., 2007; Pouso et al., 2021).
- Meet a friend for a walk, or talk to them on the phone. Research has shown that speaking to friends can boost an individual's state of mind. It gives you an opportunity to share thoughts (Hefner & Eisenberg, 2009).
- For some individuals, listening to music can help focus their mind and music has the power to soothe. Play a familiar song as you prepare for the viva (Finn & Fancourt, 2018; Wang & Agius, 2018).
- Exercise releases endorphins, which are great for giving a sense of well-being (Grossman & Sutton, 1985; Scully et al., 1998). Exercise on your own or with others, inside or outdoors. Exercise enables you to focus on something other than your thesis and viva. It offers a chance to reset your mind and thoughts. Often when someone has exercised, they feel refreshed.
- Make the most of the range of mindfulness applications available on phones and the internet. This may include apps that promote mediation. These can be listened to either on the go or at night to help ease sleeping (Armstrong & Turne, 2022; Lee et al., 2021; Schultchen et al., 2021). They have been shown to help promote good mental health.
- Listen to a podcast or an e-book—there are many available. These can help ease sleeping at night and promote good sleep (Poerio & Totterdell, 2020). Sleep deprivation can cause loss of concentration and increase irritability; in the build-up to the viva you do not want to reduce your sleep (Alkadhi et al., 2013; Borbély & Wirz-Justice, 1982). It is important that you do not forget to keep well while preparing for the viva.
- Seek out support at your HEI; the doctoral college at your HEI will have a well-being team and/counselling services if you feel you would benefit from these resources. Please contact your doctoral school for more information.
- Seek out support from your supervisory team. You may have successfully submitted your thesis, but your supervisors remain your supervisors until your viva and corrections are submitted. It is therefore important to discuss any concerns with your supervisory to help alleviate any concerns. Your supervisory teams are there to support you, so keep them aware of your state of mind.
- Keep in touch with your institution. When you hand in your thesis it can seem that you have left the institution, but you have not. It is therefore important that you tap into any resources made available to students, such

as those mentioned earlier. However, there are also other facilities to be drawn upon, such as the Chaplaincy, the well-being teams and the sports and recreation facilities, as well as the library spaces. The institution is there for you.

It is important that you acknowledge that feeling nervous about the viva is no bad thing. Sometimes, when humans are stressed and nervous, this can cause us to perform to the level of optimal alertness, behavioural and cognitive processing (Kirby et al., 2013).

Summary

- Well-being means different things to each of us.
- It is important to look after your well-being in preparation for your viva.
- Reflect on all that you have achieved in your doctoral journey.

References

Ajayi, A., & Amole, O. (2022). Open spaces and wellbeing: The impact of outdoor environments in promoting health. *Cities & Health*, 1–16.

Alkadhi, K., Zagaar, M., Alhaider, I., Salim, S., & Aleisa, A. (2013). Neurobiological consequences of sleep deprivation. *Current Neuropharmacology, 11*(3), 231–249.

Armstrong, J. W., & Turne, L. N. (2022). Mindfulness-based interventions to reduce stress and burnout in nurses: An integrative review. *British Journal of Mental Health Nursing, 11*(1), 1–11. https://doi.org/10.12968/bjmh.2020.0036

Borbély, A., & Wirz-Justice, A. (1982). Sleep, sleep deprivation and depression. *Human Neurobiology, 1*(205), 10.

Butcher, I., Morrison, R., Webb, S., Duncan, H., Balogun, O., & Shaw, R. (2022). Understanding what wellbeing means to medical and nursing staff working in paediatric intensive care: An exploratory qualitative study using appreciative inquiry. *BMJ Open, 12*(4), e056742. https://doi.org/10.1136/bmjopen-2021-056742

Finn, S., & Fancourt, D. (2018). The biological impact of listening to music in clinical and nonclinical settings: A systematic review. *Progress in Brain Research, 237*, 173–200.

Grossman, A., & Sutton, J. R. (1985). Endorphins: What are they? How are they measured? What is their role in exercise? *Medicine and Science in Sports and Exercise, 17*(1), 74–81.

Hefner, J., & Eisenberg, D. (2009). Social support and mental health among college students. *American Journal of Orthopsychiatry, 79*(4), 491–499.

Kirby, E. D., Muroy, S. E., Sun, W. G., Covarrubias, D., Leong, M. J., Barchas, L. A., & Kaufer, D. (2013). Acute stress enhances adult rat hippocampal neurogenesis and activation of newborn neurons via secreted astrocytic FGF2. *eLife, 2*, e00362.

Lee, S., Mu, C., Gonzalez, B. D., Vinci, C. E., & Small, B. J. (2021). Sleep health is associated with next-day mindful attention in healthcare workers. *Sleep Health, 7*(1), 105–112. https://doi.org/10.1016/j.sleh.2020.07.005

Peacock, J., Hine, R., & Pretty, J. (2007). The mental health benefits of green exercise activities and green care. *Report for MIND.*

Poerio, G., & Totterdell, P. (2020). The effect of fiction on the well-being of older adults: A longitudinal RCT intervention study using audiobooks. *Psychosocial Intervention, 29*(1), 29–38.

Pouso, S., Borja, Á., Fleming, L. E., Gómez-Baggethun, E., White, M. P., & Uyarra, M. C. (2021). Contact with blue-green spaces during the COVID-19 pandemic lockdown beneficial for mental health. *Science of the Total Environment, 756*, 143984.

Schultchen, D., Terhorst, Y., Holderied, T., Stach, M., Messner, E.-M., Baumeister, H., & Sander, L. B. (2021). Stay present with your phone: A systematic review and standardized rating of mindfulness apps in European app stores. *International Journal of Behavioral Medicine, 28*(5), 552–560.

Scully, D., Kremer, J., Meade, M. M., Graham, R., & Dudgeon, K. (1998). Physical exercise and psychological well being: A critical review. *British Journal of Sports Medicine, 32*(2), 111–120.

Wang, S., & Agius, M. (2018). The use of music therapy in the treatment of mental illness and the enhancement of societal wellbeing. *Psychiatria Danubina, 30*(Suppl. 7), 595–600.

'Did you read all of these pages mummy?'
'Lots of times Rory. But actually I wrote them.'
'You wrote ALL these words? WOW mummy'
A rare moment of impressing my 6 year old

Photograph by Dr Julie Menzies

Index

Printed in the United States
by Baker & Taylor Publisher Services